T0269153

CAMBRIDGE LIBRARY COLLECTION

Books of enduring scholarly value

Botany and Horticulture

Until the nineteenth century, the investigation of natural phenomena, plants and animals was considered either the preserve of elite scholars or a pastime for the leisured upper classes. As increasing academic rigour and systematisation was brought to the study of 'natural history', its subdisciplines were adopted into university curricula, and learned societies (such as the Royal Horticultural Society, founded in 1804) were established to support research in these areas. A related development was strong enthusiasm for exotic garden plants, which resulted in plant collecting expeditions to every corner of the globe, sometimes with tragic consequences. This series includes accounts of some of those expeditions, detailed reference works on the flora of different regions, and practical advice for amateur and professional gardeners.

The Gentleman Farmer

Henry Home, Lord Kames (1696–1782) gained recognition as an advocate at the Scottish bar, and subsequently as a judge whose writings on the philosophy, theory and practice of the law were hugely influential. However, he also took great interest in agriculture, and his wife's inheritance of a large estate in 1766 particularly focused his energies. The first edition of this work, published in 1776, rapidly became popular: reissued here is the enlarged second edition of 1779. Kames makes it clear that 'there never was in Scotland a period more favourable to agriculture than the present'. He begins with necessary equipment and moves on to describe the preparation of the ground, and the appropriate crops to grow for feeding to humans or cattle. This thoroughly practical work ends with an appendix in which the 'imperfection of Scotch husbandry' and a proposal for 'a board for improving agriculture' are discussed.

Cambridge University Press has long been a pioneer in the reissuing of out-of-print titles from its own backlist, producing digital reprints of books that are still sought after by scholars and students but could not be reprinted economically using traditional technology. The Cambridge Library Collection extends this activity to a wider range of books which are still of importance to researchers and professionals, either for the source material they contain, or as landmarks in the history of their academic discipline.

Drawing from the world-renowned collections in the Cambridge University Library and other partner libraries, and guided by the advice of experts in each subject area, Cambridge University Press is using state-of-the-art scanning machines in its own Printing House to capture the content of each book selected for inclusion. The files are processed to give a consistently clear, crisp image, and the books finished to the high quality standard for which the Press is recognised around the world. The latest print-on-demand technology ensures that the books will remain available indefinitely, and that orders for single or multiple copies can quickly be supplied.

The Cambridge Library Collection brings back to life books of enduring scholarly value (including out-of-copyright works originally issued by other publishers) across a wide range of disciplines in the humanities and social sciences and in science and technology.

The Gentleman Farmer

Being an Attempt to Improve Agriculture by Subjecting it to the Test of Rational Principles

Henry Home, Lord Kames

CAMBRIDGE
UNIVERSITY PRESS

University Printing House, Cambridge, CB2 8BS, United Kingdom

Cambridge University Press is part of the University of Cambridge.
It furthers the University's mission by disseminating knowledge in the pursuit of education, learning and research at the highest international levels of excellence.

www.cambridge.org
Information on this title: www.cambridge.org/9781108074377

This edition first published 1779
This digitally printed version 2015

ISBN 978-1-108-07437-7 Paperback

THE GENTLEMAN FARMER.

BEING

AN ATTEMPT TO IMPROVE

AGRICULTURE,

By ſubjecting it to the Teſt

OF

RATIONAL PRINCIPLES.

THE SECOND EDITION,
With conſiderable ADDITIONS.

Semper ego auditor tantum? JUV.

EDINBURGH:
PRINTED FOR JOHN BELL.

M, DCC, LXXIX.

EPISTLE

TO

Sir JOHN PRINGLE,

President of the ROYAL SOCIETY.

THERE are few perſons who rival Sir John Pringle in my eſteem: there are ſtill fewer whoſe friendſhip I value more. It is not, however, my purpoſe in this letter, to proclaim theſe things to the world; for what concern has the world with private connec-

tions?

tions? Ambition to have the patronage of the Royal Society to this work, is my motive for addreſſing you in this public manner. The plan it recommends, has been my guide many years; and ſucceſs has left me no doubt of its ſolidity. Your ſanction, my friend, will enſure it a gracious reception, from a body of learned men, who have diſtinguiſhed your literary merit by the greateſt honour they have to beſtow. It is my fervent deſire to be uſeful to my country: the ſtamp of that illuſtrious Society, will give a currency to the work:

every

every one will read; and every ſenſible farmer will profit by it.

Agriculture juſtly claims to be the chief of arts: it enjoys beſide the ſignal pre-eminence of combining deep philoſophy with uſeful practice. The members of your Society cannot employ their talents more profitably for their country, nor more honourably for themſelves, than in promoting and improving an art, to which Britain fundamentally is indebted for the figure it makes all the world over.

The

The theory here ſuggeſted, is in ſome meaſure new: it belongs to the Royal Society to determine, whether it be founded on ſolid principles. It will give me entire ſatisfaction, to be countenanced by a Society, which has contributed more to promote natural knowledge, than any other ſociety exiſting, or that ever did exiſt.

Your, &c.

HENRY HOME.

PREFACE.

BEHOLD another volume on hufbandry! exclaims a peevifh man on feeing the title-page: how long fhall we be peftered with fuch trite ftuff? "As long, fweet Sir, as you are "willing to pay for it: hold out your purfe, and "wares will never be wanting."

It muft indeed be acknowledged, that the commerce of books is carried on with no great degree of candour: thofe of hufbandry, with very little. A bookfeller contrives a new title, collects books upon the fubject, delivers them to his author to pick and cull; and, "Here, Sir, is a "fpick and fpan new work, full of curious mat-"ter." Agriculture is the prime of arts: every thing is made welcome on that fubject; and provided the title be new, it is to the bookfeller of no great importance, how threadbare the contents be.

Writers on agriculture, very few excepted, deliver their precepts from a ftudy lined with books, without even pretending to experience. Principles and propofitions are affumed on the

authority

authority of former writers: opinions paſs current from generation to generation; and no perſon enquires whether they wear the livery of truth. Take the following ſhort ſpecimen, drawn from a ſingle head, that of manures. Writers talk learnedly of lime, ſtone-marl, clay-marl, ſhell-marl, and of other manures; and foretel how they will operate upon ſoil, with the ſame aſſurance as if they could penetrate into their nature and eſſence. "Clay-marl," ſay they, "has more power to "deſtroy acids and to produce ſalts, than ſtone-"marl: *ergo*, a leſs quantity of it upon land is "ſufficient. Marl extracts greaſe out of woollen "cloth: greaſe is a ſpecies of oil: *ergo*, marl "extracts oil from the air. A greater quantity "of marl is laid on land, than of lime: *ergo*, it "muſt have a greater effect in attracting vege-"table food from the air. Shell-marl found un-"der moſs, is compounded of earth and the al-"kaline ſalt of rotten wood. Becauſe clay is "mixed with ſand in making brick, that mixture "muſt be an enemy to vegetation. The mixing "earth and lime with dung, makes an excellent "compoſt. Rain makes flinty ſand firmer and "more compact," &c. &c. Some of theſe propoſitions are erroneous, ſome at beſt doubtful.

And

And yet, they are but a few of many propofitions, boldly afferted by writers, though they would require the elucidation of a Newton or a Boyle. So much I will vouch for myfelf, that I have not mentioned a fingle article as certain, but what I have practifed many years with fuccefs: the inftructions contained in this book, are founded on repeated experiments and diligent obfervation. If any particular happen to be mentioned that has not come under my infpection, the reader is warned of it. In fhort, it will foon be perceived, that this is not a bookfeller's production.

The dawn of a manufacture, is irkfomely flow in its progrefs to broad day. Indocility is difficult to be overcome: habitual indolence, ftill more. Thefe obftructions vanifh in a manufacturing country. A boy learns by fight his father's trade, even before he can ply his little hand about an inftrument. He acquires activity by feeing others active: he comprehends not what it is to be idle. This obfervation is ftrongly exemplified in agriculture. Some years ago, farmers in Scotland were ignorant and indolent; nothing to be feen but weeds and trafh, not a fingle field in order. People who never faw better hufbandry, had no notion of any better. Skill in agriculture is

 fpreading

ſpreading gradually in Scotland; and young people acquire ſome knowledge by ſight, even before they think of practice. After ſuch advance, may we not hope, that our progreſs will be rapid; and that agriculture will ſoon be familiar among us, and as ſkilfully conducted as in England? May this reflection animate our landed gentry; and inflame them with a deſire, to acquire riches to themſelves, and luſtre to their country!

There never was in Scotland a period, when good leſſons in huſbandry were more ſeaſonable than at preſent. This country, growing in population, affords not corn ſufficient for its inhabitants; and yet waſte land abounds, which ſome ſkill and much induſtry would fertilize. Is it not deplorable, that in the beſt-cultivated ſhires, large patches of land ſhould produce nothing but broom and whins, not from barrenneſs but from indolence? Can greater encouragement to induſtry be wiſhed, than a ready market for every thing the ſoil produces? how different from the condition of Scotland, not more than forty years ago! Can a landholder be employed more profitably for his country, or more honourably as well as profitably for himſelf, than to rouſe emulation among

among his tenants, by kind treatment, by inſtruction, by example, and by premiums? Let him ſtudy the rules contained in this little work, all of them plain and adapted to practice. Let him convene his tenants once a-year to a hearty meal, and engage them to follow theſe rules. What if he ſhould beſtow on the deſerving, a plough or harrows of the beſt conſtruction? Land cannot be improved at a cheaper rate. It was by ſuch means, that the late John Cockburn of Ormiſtown, promoted emulation and induſtry among his people. His patriotic zeal was rewarded: he lived to ſee his eſtate in a higher degree of cultivation, than even to this day is ſeen in any other part of Scotland. The ſame means were employed more extenſively, by the late Earl of Findlater: the ſkill and perſeverance of that nobleman, raiſed his tenants from a torpid ſtate, to a ſurpriſing degree of activity; and few can now vie with them, either for induſtry or knowledge. Had other landlords been equally active, how flouriſhing would agriculture have been in Scotland? How great a change to our advantage would there have been in the commercial balance, had we been feeding our neighbours inſtead of being fed by them; had we been in the courſe

courſe of receiving money for our corn, inſtead of receiving corn for our money! The field is ſtill open: let us join hearts and hands to redeem time wofully miſpent. I ſay again, there never was in Scotland a period more favourable to agriculture than the preſent.

Agriculture is a very ancient art. It has been practiſed every where without intermiſſion; but with very little attention to principles. In ſtudying the principles laid down by writers, I found myſelf in a ſort of labyrinth, carried to and fro without any certain direction. After a long courſe of reading, where there was nothing but darkneſs and diſcrepance, I laid aſide my books, took heart, and like Des Cartes commenced my inquiries with doubting of every thing. I reſorted to the book of nature: I ſtudied it with attention: and the ſecond part of this work contains the reſult of my inquiries. It is far from my thoughts, to impoſe my opinions upon others: I pretend only to have reduced the theory of agriculture into a ſort of ſyſtem, more conciſe at leaſt, and more conſiſtent, than has been done by other writers. Many eyes are better than one; and if my theory ſhall be found erroneous, the many that have erred will ſerve in ſome meaſure

ſure to keep me in countenance. I am not however afraid of any groſs error. An *imprimatur* from one of the ableſt chemiſts of the preſent age, has given me ſome confidence of being in the right tract *.

I have all along ſtudied brevity, as far as conſiſtent with perſpicuity; and therefore, have confined myſelf to matters that I know to be of real uſe in practice. I am ambitious to have my plan followed, becauſe ſucceſsful experience has proved it to be ſolid: but I ſhould not hope for many readers, if I hazarded the tiring them with unneceſſary matters. Varro *de re ruſtica*, appears to be very ſparing of inſtruction; but rivals Ariſtotle himſelf in the ſubtilety of his diviſions. "Nunc "dicam agri quibus rebus colantur. Quas res "alii dividunt in duas partes, in homines et admi-"nicula hominum: ſine quibus rebus colere non "poſſunt. Alii in tres partes, inſtrumenti genus "vocale, et ſemivocale, et mutum. Vocale, in "quo ſunt ſervi: ſemivocale, in quo ſunt bo-"ves: mutum, in quo ſunt plauſtra." Such puerile diviſions, may be of uſe to ſwell a volume; but give no inſtruction, and are extremely tireſome.

* Dr Black profeſſor of chemiſtry in the college of Edinburgh.

I cannot finiſh this preface without warmly recommending agriculture to gentlemen of land-eſtates; for whoſe uſe chiefly this work is intended. In every well-governed ſtate, agriculture has been duly honoured. In ancient Perſia, a feſtival was yearly celebrated, in which huſbandmen were freely admitted to the King's table. "From your labours, ſaid the King, we receive "our ſuſtenance; and by us you are protected. "Being mutually neceſſary to each other, let us, "like brethren, live together in amity." The great emperor of China, performs yearly the ceremony of holding the plough, to ſhow that no man is above being a farmer. The iſland Miletus, during many years, had been afflicted with factions: the government was ſettled by ſome wiſe men of Paros, a neighbouring iſland. Theſe men having ſurveyed the iſland, and marked the poſſeſſor of every well-cultivated farm, convocated an aſſembly of the people, and appointed theſe perſons to be governors. "The perſon," ſaid they, "who governs his private affairs with prudence and induſtry, is qualified "to govern thoſe of the public." The king of Tunis, invaded by a powerful enemy, promiſed to a neighbour who aſſiſted him, the philoſopher's pher's ſtone. He ſent a plough; terming it the philoſopher's ſtone, becauſe it would produce rich crops, to procure gold in plenty.

In the view of profit, agriculture is fit for every man. In the view of pleaſure, it is of all occupations the beſt adapted to gentlemen in a private ſtation. Matter crouds upon me, and I am at a loſs where to begin. Agriculture correſponds to that degree of exerciſe, which is the beſt preſervative of health. It requires no hurtful fatigue, on the one hand, nor indulges, on the other, indolence, ſtill more hurtful. During a throng of work, the diligent farmer will ſometimes be early and late in the field: but this is no hardſhip upon an active ſpirit. At other times, a gentleman who conducts his affairs properly, may have hours every day, to beſtow on reading, on his family, on his friends.

Agriculture is equally ſalutary to the mind. In the management of a farm, conſtant attention is required to the ſoil, to the ſeaſon, and to different operations. A gentleman thus occupied, becomes daily more active, and is daily gathering knowledge: as his mind is never ſuffered to lanquiſh, he is ſecure againſt the diſeaſe of low ſpirits.

But

But what I chiefly inſiſt on is, that laying aſide irregular appetites and ambitious views, agriculture is of all occupations the moſt conſonant to our nature; and the moſt productive of contentment, the ſweeteſt ſort of happineſs. In the firſt place, it requires that moderate degree of exerciſe, which correſponds the moſt to the ordinary ſucceſſion of our perceptions. Fox-hunting produces a ſucceſſion too rapid: angling produces a ſucceſſion too ſlow. Agriculture correſponds not only more to the ordinary ſucceſſion, but has the following ſignal property, that a farmer can direct his operations, with that degree of quickneſs and variety, which is agreeable to his own train of perceptions. In the next place, to every occupation that can give a laſting reliſh, hope and fear are eſſential. A fowler has little enjoyment in his gun, who miſſes frequently; and he loſes all enjoyment, when every ſhot is death: a poacher, ſo dextrous, may have pleaſure in the profit, but none in the art. The hopes and fears that attend agriculture, keep the mind always awake, and in an enlivening degree of agitation. Hope never approaches certainty ſo near, as to produce ſecurity; nor is fear ever ſo great, as to create deep anxiety and diſtreſs. Hence it is, that a

gentleman

gentleman farmer, tolerably ſkilful, never tires of his work; but is as keen the laſt moment as the firſt. Can any other employment compare with farming in that reſpect? In the third place, no other occupation rivals agriculture, in connecting private intereſt with that of the public. How pleaſing to think, that every ſtep a man makes for his own good, promotes that of his country! Even where the balance happens to turn againſt the farmer, he has ſtill the comfort that his country profits by him. Every gentleman farmer muſt of courſe be a patriot; for patriotiſm, like other virtues, is improved and fortified by exerciſe. In fact, if there be any remaining patriotiſm in a nation, it is found among that claſs of men.

A gentleman farmer who is diſpoſed to embelliſh his fields, has a great advantage over others. He can execute that pleaſing work at the cheapeſt rate, by employing upon it his farm ſervants and cattle, every vacant hour. This ſlow method is indeed ill ſuited to the ardour of an Indian Nabob, impatient for enjoyment. But is not the advantage clearly on the ſide of the farmer? The refined pleaſure of embelliſhments, ariſes from a ſlow progreſs; which affords leiſure to feaſt the eye upon every new production.

In former times, hunting was the only businefs of a gentleman. The practice of blood made him rough and hard-hearted: he led the life of a dog, or of a favage; violently active in the field, fupinely indolent at home. His train of ideas was confined to dogs, horfes, hares, foxes: not a rational idea entered the train, not a fpark of patriotifm, nothing done for the public, his dependents enflaved and not fed, no hufbandry, no embellifhment, loathfome weeds round his dwelling, diforder and dirt within. Confider the prefent mode of living. How delightful the change, from the hunter to the farmer, from the deftroyer of animals to the feeder of men! Our gentlemen who live in the country, have become active and induftrious. They embellifh their fields, improve their lands, and give bread to thoufands. Every new day promotes health and fpirits; and every new day brings variety of enjoyment. They are happy at home; and they wifh happinefs to all.

As the fcene of my experience has all along been in Scotland, my native country, I am fhy to recommend this plan of hufbandry to any but to my countrymen. I have, however, a thorough conviction, that giving allowance for flight variations

variations of climate, the plan will suit England, France, Italy, and every other country situated within either of the temperate zones.

Among the old Romans there were excellent writers on husbandry; but I cannot prevail on myself to think that their practice was answerable. They were enslaved by observing superstitiously omens, prognostics, unlucky days, *&c.* which frequently prevented them from taking advantage even of the most favourable weather. They were conducted in a great measure by chance; and little scope was left for skill or foresight. Examples may be found in every one of their writers on husbandry. I shall confine myself to Columella the most celebrated of them. He gives the following receipt against the wevil from his own experience. "At the change of "the moon, pull your beans before day-light: "when perfectly dried before full moon, thresh "them; and the seeds laid up in a granery, "will suffer no damage from the wevil." He forbids vetches to be sowed before the twenty-fifth day of the moon; that otherwise they will be hurt by the snail. His way to prevent rats and mice from preying on a vineyard, is to prune the vines in the night-time when the moon is full.

full. The ſeed of medic, ſays he, ought to be covered with a wooden rake; for that iron is deſtructive to it. He orders frequent digging about a tree new planted; but diſcharges the ground to be touched with an iron tool after the planting. He quotes Ariſtotle as his authority, that among ſheep the way to procreate a male is to admit the ram when the north wind blows; and to admit the ram when the ſouth wind blows, in order to procreate a female. Tie up before copulation the left teſticle, and the ſtallion will produce a male: tie up the right teſticle, and he will produce a female. What does the reader think of the following prognoſtic? If a horſe after covering a mare deſcend on the right hand, the foal will be a male: if on the left, it will be a female. It was believed by the Roman writers without a ſingle exception, that mares in Luſitania were impregnated by the weſt wind. I congratulate my countrymen for their happy deliverance from ſuch heavy fetters. There is now a fair field for exerciſing our talents, natural and acquired; and if we fail in any article, we have ourſelves only to blame, not deſtiny.

In arts and ſciences, a plentiful ſource of obſcurity and indiſtinctneſs, is the uſing a word in different

different ſenſes, without warning the reader of it. Conſidering what volumes have been compoſed on agriculture, it is amazing how little preciſion there is in the terms of art. Take the following inſtance. The word *furrow* is employed to ſignify not only the hollow made by the plough, but the earth taken out of that hollow, and alſo the hollow between ridges. Better coin words than write indiſtinctly. Let *furrow* be appropriated to the ſpace in which the plough moves, and alſo to the hollow between ridges; which will not occaſion any confuſion. But I venture to diſtinguiſh the earth moved by the plough, by the name of the *furrow-ſlice*. The ſmall hollows that appear between the ſlices when a ridge is plowed, may be termed *ſeams*.

Earth, land, ground, ſoil, are not ſynonimous; and therefore, in correct writing, their meaning ought to be aſcertained. Earth is oppoſed to metals, foſſils, and ſuch like. Land is any indefinite ſection of this globe *a cœlo ad centrum*. Ground is the ſurface of land; and every quality of a ſurface can be attributed to it, hilly ground, flat ground, ſmooth ground, rough ground, ſoft ground, hard ground. Quantity is an attribute of land, improperly of ground. We ſay current-

ly

ly a quantity of land, not a quantity of ground otherways than figuratively. The earth we tread on, with reſpect to its power of nouriſhing plants, is termed ſoil, rich ſoil, poor ſoil, dry ſoil, wet ſoil, clay ſoil, ſandy ſoil. Staple is uſed by Engliſh writers with reſpect to the nature of the ſoil, In common language theſe terms are not diſtinctly ſeparated; nor do I pretend that my definitions are altogether accurate. To fix a preciſe meaning to each will probably require a century or two more.

SCOTCH and ENGLISH Meaſures and Weights compared.

The meaſure of oats and barley is the ſame. The meaſure of wheat, peaſe, and beans, the ſame, both Linlithgow meaſure.

The wheat firlot of Scotland contains
of cubic inches - - $2197\frac{34}{100}$
The barley firlot contains - $3205\frac{54}{100}$
The Wincheſter buſhel of England $2150\frac{42}{100}$

Therefore four firlots of barley are nearly equal to ſix firlots of wheat. And the Wincheſter buſhel nearly equal to our wheat firlot.

The

The Scotch acre contains 55,353$\frac{6}{10}$ ſquare feet.
The Engliſh acre contains 43,360 ſquare feet.

Therefore four Scots acres are little leſs than five Engliſh.

WEIGHTS.

The Troy ounce is -	480 Troy grains
The Averdupois ounce is	437$\frac{1}{2}$
The Scotch ounce is	476

100 pounds net Amſterdam weight is equal to 108$\frac{2}{3}$ pounds Averdupois.

CON-

CONTENTS.

PART I.

Practice of Agriculture.

CHAP. I.

Instruments of Husbandry.

2. *Dry*

CHAP. X.

CHAP. XI.

CHAP. XII.

CHAP. XIII.

CHAP. XIV.

PART II.

CHAP. I.

CHAP. II.

CHAP. III.

APPENDIX.

THE

THE GENTLEMAN FARMER.

NATURAL HISTORY is confined to effects, leaving causes to Natural Philosophy. From a number of effects, Natural Philosophy ascends by induction to the immediate cause; and many of these causes are by another induction found to proceed from one more general and comprehensive. Such is the mode of reasoning in Natural Philosophy, till we arrive at an ultimate cause; that is, a cause beyond which we cannot penetrate. Most writers treat husbandry as a branch of Natural History. Some, more bold, consider it as also a branch of Natural Philosophy: they begin with effects, and endeavour to unfold the causes or principles. In addressing this treatise to Gentlemen, I attempt both. This suggests a division into two parts. In the firſt part, which indeed is the moſt uſeful, the beſt practice in every branch of huſbandry is carefully explained. In the ſecond, with timid ſteps and ſlow, I endeavour to trace out a few cauſes or principles that have an immediate influence upon practice.

Plate 1

Chain Plough

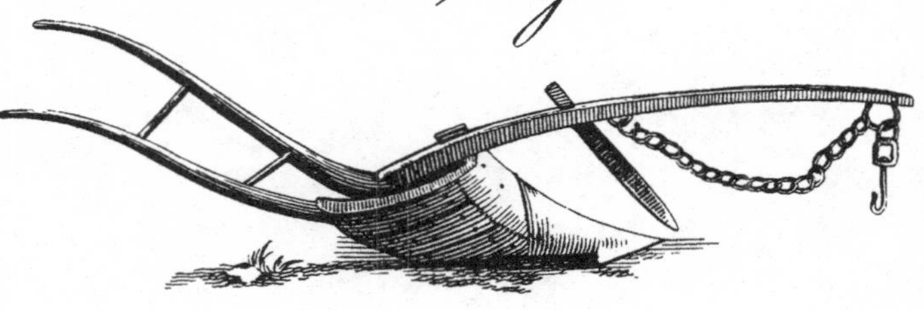

Brake

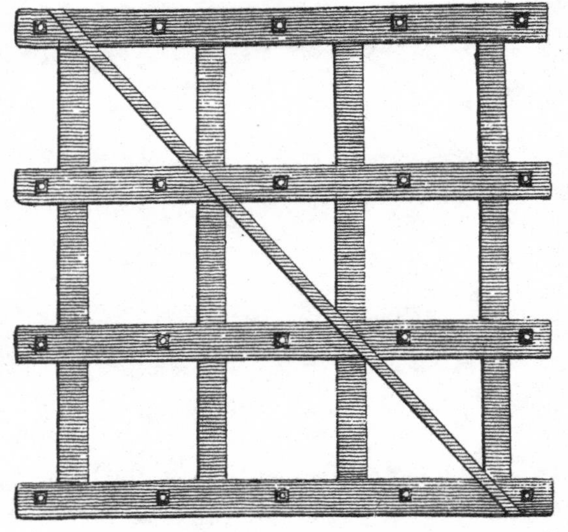

Models of the Instruments here delineated are ready for public inspection at the work yard of Mr. Crichton Coachmaker Edinr.

PART I.

PRACTICE of AGRICULTURE.

IT is unneceſſary here to make a liſt of what is contained in this part. The particulars are in the prefixed table.

CHAPTER I.

INSTRUMENTS of HUSBANDRY.

IN the natural courſe of ideas, the hand goes before the inſtrument, and the inſtrument before the operation. But as the nature of man is a ſubject too important and too extenſive to be ranged under any other head, I begin this treatiſe of agriculture with the inſtruments employed.

1. The PLOUGH.

THIS is the moſt uſeful inſtrument ever was invented. It is of more uſe than even the ſpinning-wheel: for men may make a ſhift for cloathing without that inſtrument, but a country cannot be populous without the plough.

The only plough uſed in Scotland, till of late, is a ſtrong heavy inſtrument, about thirteen feet from

from the handles to the extremity of the beam, and commonly above four feet from the back end of the head to the point of the sock. It is termed the *Scotch plough*, to distinguish it from other forms; and it needs no particular description, as it may be seen in every field. It may well be termed the *Scotch plough*; for of all forms it is the fittest for breaking up stiff and rough ground, especially where stones abound; and no less fit for strong clays hardened by drought. The length of its head gives it a firm hold of the ground: its weight prevents it from being thrown out by stones: the length of the handles gives the ploughman great command to direct its motion: and by the length of its head, and of its mouldboard, it lays the furrow-slice cleverly over. The Scotch plough was contrived during the infancy of agriculture, and was well contrived: in the soils above described, it has not an equal.

But in tender soil it is improper, because it adds greatly to the expence of plowing, without any counterbalancing benefit. By the length of the head and mouldboard the friction is encreased, requiring a greater number of oxen or horses than are necessary in a shorter plough. There is in its form, another particular that resists the draught: the mouldboard makes an angle with the sock, instead of making a line with it gently curving backward. An objection against it still more material, is, that is does not stir

the

the ground perfectly: the hinder part of the wrift rifes a foot above the fole of the head; and the earth that lies immediately below that hinder part, is not fufficiently turned over. This is ribbing land below the furface, fimilar to what is done by ignorant farmers on the furface.

Thefe defects muft be fubmitted to in a foil that requires a ftrong heavy plough; but may be avoided in a cultivated foil by a plough differently conftructed. Of all the ploughs fitted for a cultivated foil free of ftones, I boldly recommend a plough introduced into Scotland about twelve years ago, by James Small in Blackadder Mount, Berwickfhire; which is now in great requeft; and with great reafon, as it avoids all the defects of the Scotch plough. The fhortnefs of its head and of its mouldboard leffen the friction greatly: from the point of the fock to the back part of the head it is only thirty inches; and the whole length, from the point of the beam to the end of the handles, between eight and nine feet. The fock and mouldboard make one line gently curving; and confequently gather no earth. Inftead of a wrift, the under edge of the mouldboard is in one plain with the fole of the head; which makes a wide furrow, without leaving any part unftirred. It is termed the *chain-plough*, becaufe it is drawn by an iron chain fixed to the back part of the beam immediately before the coulter. This has two advantages: firft, by means of of a muzzle, it makes the plough go deep or ſhallow; and next, it ſtreſſes the beam leſs than if fixed to the point, and therefore a ſlenderer beam is ſufficient. Theſe particulars will be better underſtood from inſpecting the annexed figure.

This plough may well be conſidered as a capital improvement; not only by ſaving expence, but by making better work. It is proper for loams, for carſe clays, and, in general, for every ſort of tender ſoil free of ſtones. It is even proper for opening up paſture-ground that has formerly been well cultivated.

To finiſh an account of the plough, I muſt add a word about the ſock. A ſpiked ſock is uſed in the Scotch plough, and is eſſential in ſtony land. But a feathered ſock ought always to be uſed in tender ſoil, free of ſtones: it cuts the earth in the furrow, and makes neat work. It is indiſpenſable in ground where roots abound, as it cuts them below the ſurface, and prevents their growing. I eſteem the feathered ſock to be a valuable improvement. The induſtrious farmer would even borrow money to clear his ground of ſtones, in order to introduce it: in a twenty-years leaſe, the profit of it would repay the expence tenfold.

A wheel-plough poſſeſſes one advantage, that it requires no ſkill in the ploughman. But it requires more ſtrength of cattle, the friction of the wheels being added to that of the ground. But

a

a much greater defect is, that the furrows muſt partake of the inequalities of the ſurface: every ſtone, every clod, diſturbs its motion. To croſs with a wheel-plough a field of high and narrow ridges, ſuch as is fit for turnip, would ridge the furrows like the ſurface, and retain every drop of rain that falls on it. Therefore, I greatly prefer a plough without wheels, in expert hands.

Some ploughs are made with two ſmall wheels running in the furrow, in order to take off the friction of the head; and this plough is recommended in a book, intitled, *The complete Farmer.* But all complicated ploughs are baubles; and this as much as any. The pivots of ſuch wheels are always going wrong; and beſide, they are choked ſo with earth, as to increaſe the friction inſtead of diminiſhing it.

If we look back thirty years, ploughs of different conſtructions did not enter even into a dream. The Scotch plough was univerſally uſed; and no other was known. There was no leſs ignorance as to the number of cattle neceſſary for this plough. In the ſouth of Scotland, ſix oxen and two horſes were univerſal; and in the north, ten oxen, ſometimes twelve. The firſt attempt to leſſen the number of oxen, was in Berwickſhire. The low part of that county abounds with ſtone and clay marl, the moſt ſubſtantial of all manures, which had been long uſed by one

or

or two gentlemen. About twenty-five years ago it acquired reputation, and ſpread rapidly. As two horſes and two oxen were employed in every marl-cart; the farmer, in ſummer-fallowing and in preparing land for marl, was confined to four oxen and two horſes. And as that manure afforded plenty of ſucculent ſtraw for oxen, the farmer was ſurpriſed to find, that four oxen did better now than ſix formerly. Marling, however, a laborious work, proceeded ſlowly, till people were taught by a noted farmer in that country, what induſtry can perform by means of power properly applied. It was reckoned a mighty taſk to marl five or ſix acres in a year. That gentleman, by plenty of red clover for his working-cattle, accompliſhed the marling fifty acres in a ſummer, once fifty-four. Having ſo much occaſion for oxen, he tried with ſucceſs two oxen and two horſes in a plough; and that practice became general in Berwickſhire.

Now here appears with luſtre the advantage of the chain-plough. The great friction occaſioned in the Scotch plough by a long head, and by the angle it makes with the mouldboard, neceſſarily requires two oxen and two horſes, whatever the ſoil be. The friction is ſo much leſs in the chain-plough, that two good horſes are found ſufficient in every ſoil that is proper for it. And as good luck ſeldom comes alone more than bad, the reducing the draught to a couple of horſes has

has another advantage, that of rendering a driver unneceſſary; no ſlight ſaving at preſent, where a ſervant's wages and maintenance are very ſmart articles. This ſaving on every plough, where two horſes and two oxen were formerly uſed, will by the ſtricteſt computation be fifteen pounds Sterling yearly; and where four horſes were uſed, no leſs than twenty pounds Sterling. There is now ſcarce to be ſeen in the low country of Berwickſhire a plough with more than two horſes; which undoubtedly in time will become general. Had the practice of four horſes in a plough continued, in vain would one have expected in this country a good breed of labouring horſes, when four of our own paultry kind were more than ſufficient: But, by better dreſſing, plowing became an eaſier work, and two horſes in many ſoils were found ſufficient. This diſcovery became gradually more general by lighter ploughs and ſtouter horſes. There is now a demand for a better breed of labouring horſes; which probably in time will perfect the breed. I know but of one further improvement, that of uſing two oxen inſtead of two horſes. That draught has been employed with ſucceſs in ſeveral places; and the ſaving is ſo great, that it muſt force its way every where. I boldly affirm, that no ſoil ſtirred in a proper ſeaſon, can ever require more than two horſes and two oxen in a plough, even ſuppoſing the ſtiffeſt clay. In all other ſoils, two good horſes or two good oxen, abreaſt, may be relied on for every operation of the chain-plough.

A chain-plough of a ſmaller ſize than ordinary, drawn by a ſingle horſe, is of all the moſt proper for horſe-hoeing, ſuppoſing the land to be mellow, which it ought to be for that operation. It is ſufficient for making furrows to receive the dung, for plowing the drills after dunging, and for hoeing the crop.

A ſtill ſmaller plough of the ſame kind, I warmly recommend for a kitchen-garden. It can be reduced to the ſmalleſt ſize, by being made of iron; and where the land is properly dreſſed for a kitchen-garden, an iron plough drawn by a horſe of the ſmalleſt ſize will ſave much ſpade-work. Strange is the effect of cuſtom without thought! Thirty years ago, a kitchen-garden was an article of luxury merely, becauſe at that time there could be no cheaper food than oat-meal. At preſent the farmer maintains his ſervants at double expence, as the price of oat-meal is doubled; and yet he has no notion of a kitchen-garden, more than he had thirty years ago. He never thinks, that living partly on cabbage, kail, turnip, carrot, would ſave much oat-meal: nor does he ever think, that change of food is more wholeſome than vegetables alone, or oat-meal alone. I need not recommend potatoes, which in our late ſcanty crops of of corn have proved a great bleſſing: without them the labouring poor would frequently have been reduced to a ſtarving condition. Would the farmer but cultivate his kitchen-garden with as much induſtry as he beſtows on his potato-crop, he needed never fear want; and he can cultivate it with the iron plough at a very ſmall expence. It may be held by a boy of twelve or thirteen; and would be a proper education for a ploughman. But it is the landlord who ought to give a beginning to the improvement. A very ſmall expence would encloſe an acre for a kitchen-garden to each of his tenants; and it would excite their induſtry to beſtow an iron plough on thoſe who do beſt.

Nor is this the only caſe where a ſingle-horſe plough may be profitably employ'd. It is ſufficient for ſeed-furrowing barley, where the land is light and well dreſſed. It may be uſed in the ſecond or third plowing of fallow, to encourage annual weeds, which are deſtroy'd by ſubſequent plowings.

To procure food is indeed the chief object of the plough, but not its only object. Good roads are eſſential to internal commerce; and the expence of making them may be conſiderably leſſened by the plough. As this hitherto has been little thought of, an explanation is neceſſary. The method in uſe is, to form a road with the pick-axe, the ſpade, and the wheelbarrow. Even where

where a pick-axe is not neceſſary, you ſee ten or twelve men preſſing down the ſpade with the foot oftner than once before a ſufficient load of earth can be raiſed; — dearly bought by the workmen, and ſtill more dearly by the employer. Where a pick-axe is neceſſary, there muſt be a great addition of hands; for ten pickmen are no more than ſufficient to looſen what can be thrown up with four or five ſpades. Now a great part of this labour may be ſaved by the plough. The Scotch plough, fortified with iron plates, and the head connected with the beam by a bar of iron, is an excellent inſtrument for making roads. Suppoſe a new road is purpoſed thirty feet wide, plough it up into a ridge, beginning in the middle; and plough it a ſecond time in the ſame manner. Where the ground is ſoft, and requires to be raiſed high, a very deep furrow is neceſſary. Where the ground is firm, a ſhallow furrow is ſufficient. After theſe two plowings are finiſhed, if the ſides of the road be too ſteep, leave ſix feet in the middle, and go round the remainder in a third plowing, gathering it toward the top. If the ſides be ſtill too ſteep, leave twelve feet in the middle, and gather up the remainder as in the former plowing. If theſe operations be well conducted, the water-channels on each ſide of the road will be two feet lower than the ſurface of the adjacent ground. Smooth the road with a drag-harrow; and correct with a ſpade any

remaining

remaining defects or inequalities, which is a very easy work. Thus the road is completely formed to receive a covering of gravel, or of stones beat small.

A plough may also be used advantageously in making ditches for enclosing. The immense cost of loosening a hard or stony soil by the pick-axe and spade, may be totally saved by the plough. The surface-earth is commonly soft: after it is removed with the spade, let a plough, drawn by three horses in a line, go round and round the space intended for the ditch, cleaving it as if it were a ridge. After the earth thus loosened is thrown up with the shovel, renew the plowing and shovelling till you come within eight inches of the bottom; and to these eight inches apply the pick-axe and spade. One precaution is necessary, that no more be plowed at a time than can be thrown up the same day. If rain fall in any quantity, the ground tilled will become mud, very improper to be laid upon thorns. In this operation there is no occasion for the coulter: it is rather an impediment. I esteem this a valuable discovery for Scotland; which being more pestered with high winds than England, requires the more to be enclosed. The expence of enclosing with hedge and ditch the ordinary way, is great; and the ditch is the most expensive part. Two-thirds of the expence may be saved by the plough, in hard ground.

In

In every caſe where earth is to be removed, the plough is uſeful; as for example in a gravel-pit opened for high-roads. The gravel may be ſo looſened by the plough as to require a ſhovel only for filling it into the carts.

Has any one ſtumbled on the thought of uſing the plough in planting young trees? The method I have practiſed, is to mark out lines due north and ſouth, at intervals of ten or twelve feet. Let three deep furrows be made with the plough at the ſide of each line. Lay the ſod of the eaſt-moſt furrow upon the other two, which will raiſe a ſcreen about two feet high. Plant along the furrow from whence the ſod was removed, and the ſcreen behind will make good ſhelter. This method is chiefly intended for firs in a bare moor. Before the firs riſe much above the ſcreen, the roots will have taken ſuch hold of the ground as to reſiſt even weſterly winds: ſcarce a plant will fail, if they be wholeſome. Three thouſand firs planted in this manner may be ſufficient for an acre, equal to five or ſix thouſand in the ordinary way.

A fir makes a choice nurſe for other trees. After three years, even in the pooreſt ſoil, the firs begin to grow with vigour; and then is the time for planting among them oaks, elms, or other trees; cutting down the firs from time to time to make room for theſe trees. Thus, the method here pointed out for planting firs, is the beſt preparation for raiſing all other barren trees.

2. The

2. The BRAKE, or DRAG-HARROW.

THE brake is a large and weighty harrow, the purpofe of which is to reduce a ftubborn foil, where an ordinary harrow makes little impreffion. It confifts of fquare bulls*, four in number, each fide five inches, and fix feet and a half in length. The teeth are feventeen inches long, bending forward like a coulter. Four of them are inferted in each bull, fixed above with a fcrew-nut, having twelve inches free below, with a heel clofe to the under part of the bull, to prevent it from being pufhed back by ftones. The nut above makes it eafy to be taken out for fharping. This brake requires four horfes or four oxen. One of a leffer fize will not fully anfwer the purpofe: one of a larger fize will require fix oxen; in which cafe the work may be performed at lefs expence with the plough. See the figure annexed.

This inftrument may be applied to great advantage in the following circumftances. In fallowing ftrong clay that requires frequent plowings, a brakeing between every plowing, tends to pulverize the foil, and to render the fubfequent plowings more eafy. In the month of March or

* The wood of a brake, or of a harrow in which the teeth are inferted, is termed in Scotland a *bull*.

or April, when ſtrong ground is plowed for barley, eſpecially if bound with couch-graſs, a croſs-brakeing is preferable to a croſs-plowing, and is done at half the expence. When ground is plowed from the ſtate of nature, and after a competent time is croſs-plowed, the brake is applied with great ſucceſs immediately after, to reduce the whole to proper tilth.

Let it be obſerved, that a brake with a greater number of teeth than above mentioned, is improper for ground that is bound together with the roots of plants; which is always the caſe of ground new broken up from its natural ſtate. The brake is ſoon choked, and can do no execution till freed from the earth it holds. A leſs number of teeth would be deficient in pulverizing the ſoil.

To ſet in a clear light the advantages of this inſtrument, we ſhall ſtop a little, to obſerve how inſufficient the common harrow is for any of the operations mentioned. It may anſwer for covering the ſeed, and may do tolerably well in light and free ſoil; but is altogether inſufficient for reducing ſtiff ſoil. The harrow with wooden teeth is a ridiculous inſtrument, fit to raiſe laughter inſtead of raiſing mould. The poor farmer labours with it, thinks he is doing an uſeful work, when all the time he is doing nothing. It ought to be prohibited by the landlord; for a tenant with ſuch an inſtrument cannot pay a rent that the

Plate 2.

First Harrow

Second Harrow

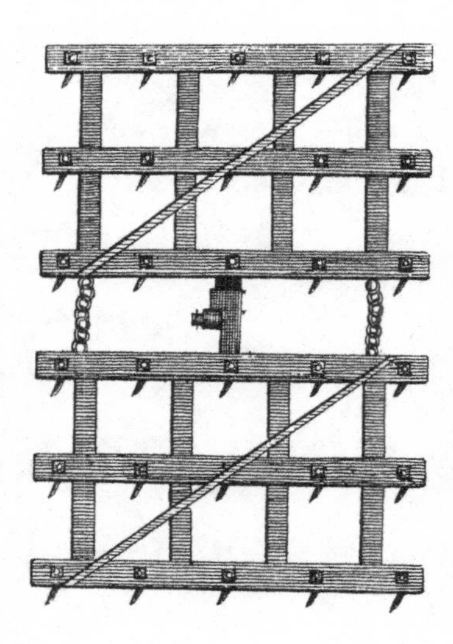

the farm properly cultivated will eafily bear. Tho' the brake has been known above twenty years, yet none but gentlemen, and a few felect tenants, have ever thought of it: in fome counties even the name remains unknown. It belongs to gentlemen of fortune, for their own intereft, to make it more general. The neceffity of fome inftrument, more effectual than the common harrow, for reducing a ftubborn foil, has led farmers to put three or four harrows, one above another, in order to prefs the undermoft into the ground. This fubftitute to the brake is far inferior in its effect; befide, that the undermoft harrow is torn to pieces in an inftant. To conclude this article, a farmer who has no brake, wants a capital inftrument of hufbandry. Its price above that of common harrows, bears no proportion to the profit.

3. The HARROW.

HARROWS are commonly confidered as of no ufe but to cover the feed. They have another ufe fcarce lefs effential, which is to prepare land for the feed. This is an article of importance for producing a good crop. And to fhew how imperfectly either of thefe purpofes is performed by the common harrow, take the following account of it.

The harrow commonly uſed is of different forms. The firſt I ſhall mention has two bulls, four feet long and eighteen inches aſunder, with four wooden teeth in each. A ſecond has three bulls and twelve wooden teeth. A third has four bulls, and twenty teeth, of wood or iron, ten, eleven or twelve inches aſunder. Now, in fine mould, the laſt may be ſufficient for covering the ſeed; but none of them are ſufficient to prepare for the ſeed any ground that requires ſubduing. The only tolerable form is that with iron teeth; and the bare deſcription of its imperfections, will ſhew the neceſſity of a more perfect form. In the firſt place, this harrow is by far too light for ground new taken up from the ſtate of nature, for clays hardened with ſpring-drought, or for other ſtubborn ſoils: it floats on the ſurface, and after frequent returns in the ſame track, does nothing effectually. In the next place, the teeth are too thick ſet, by which the harrow is apt to be choked, eſpecially where the earth is bound with roots, which is commonly the caſe. At the ſame time, the lightneſs and number of teeth keep the harrow upon the ſurface, and prevent one of its capital purpoſes, that of dividing the ſoil. Nor will fewer teeth anſwer for covering the ſeed properly. In the third place, the teeth are too ſhort for reducing a coarſe ſoil to proper tilth; and yet it would be in vain to make them longer, becauſe the harrow is too light for going deep

deep into the ground. Further, the common harrows are ſo ill conſtructed, as to ride at every turn one upon another. Much time is loſt in diſengaging them. What pity it is, that an induſtrious farmer ſhould be reduced to ſuch an imperfect inſtrument, which is neither fit to prepare the ground for ſeed, nor to cover it properly. And I now add, that it is equally unfit for extirpating weeds. The ground is frequently ſo bound with couch-graſs, as to make the furrow-ſlice ſtand upright, as when old lea is plowed: notwithſtanding much labour, the graſs-roots keep the field, and gain the victory. What follows? The farmer at laſt is reduced to the neceſſity of leaving the weeds in peaceable poſſeſſion, becauſe his field will no longer bear corn.

A little reflection, even without experience, will make it evident, that the ſame harrows, whatever be the form, can never anſwer all the different purpoſes of harrowing, nor can operate equally in all different ſoils, rough or ſmooth, firm or looſe. Looking back not many years above thirty, no farmer in Scotland had the ſlighteſt notion of different ploughs for different purpoſes. The Scotch plough was the only one known. Different ploughs are now introduced; and it is full time to think of different harrows. Rejecting the common harrows, as in every reſpect inſufficient, I boldly recommend the following. I uſe three of them of different forms, for different

different purpoſes. They are all of the ſame weight, drawn each by two horſes. Birch is the beſt wood for them, becauſe it is cheap, and not apt to ſplit. The firſt is compoſed of four bulls, each four feet ten inches long, three and a quarter inches broad, and three and a half deep; the interval between the bulls eleven and three fourths inches; ſo that the breadth of the whole harrow is four feet. The bulls are connected by four croſs-bars, which go through each bull, and are fixed by wooden nails driven through both. In each bull five teeth are inſerted, ten inches free under the bull, and ten inches aſunder. They are of the ſame form with thoſe of the brake, and inſerted into the wood in the ſame manner. Each of theſe teeth is three pounds weight; and where the harrow is made of birch, the weight of the whole is ſix ſtone fourteen pounds Dutch. An erect bridle is fixed at a corner of the harrow, three inches high, with four notches for drawing higher or lower. To this bridle a double tree is fixed for two horſes drawing abreaſt, as in a plough. And to ſtrengthen the harrow, a flat rod of iron is nailed upon the harrow from corner to corner in the line of the draught.

The ſecond harrow conſiſts of two parts, connected together by a crank or hinge in the middle, and two chains of equal length, one at each end, which keep the two parts always parallel, and at the ſame diſtance from each other. The crank

Third Harrow

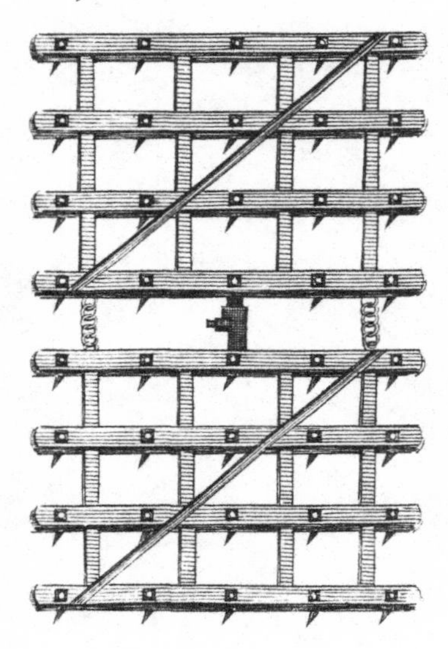

Cleaning Harrow

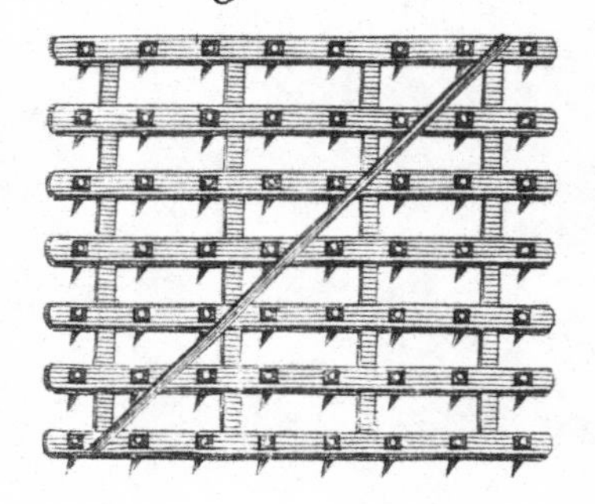

Grass seed Harrow

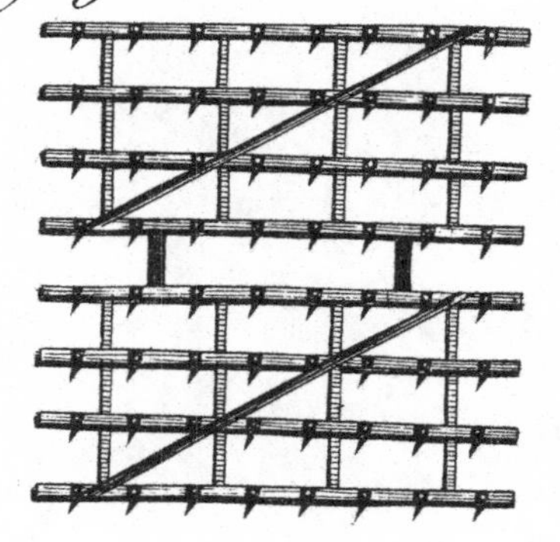

crank is ſo contrived, as to allow the two parts to ply to the ground like two unconnected harrows; but neither of them to riſe above the other, more than if they were a ſingle harrow without a joint. In a word, they may form an angle downward, but not upward. Thus they have the effect of two harrows in curved ground, and of one weighty harrow in a plain. This harrow is compoſed of ſix bulls, each four feet long, three inches broad, and three and a half deep. The interval between the bulls nine and a half inches; which makes the breadth of the whole harrow, including the length of the crank, to be five feet five inches. Each bull has five teeth, nine inches free under the wood, and ten inches aſunder. The weight of each tooth is two pounds; the reſt as in the former.

The third conſiſts alſo of two parts, connected together like that laſt mentioned. It has eight bulls, each four feet long, two and a half inches broad, and three deep. The interval between the bulls is eight inches; and the breadth of the whole harrow, including the length of the crank, is ſix feet four inches. In each bull are inſerted five teeth, ſeven inches free under the wood, and ten and a half inches aſunder, each tooth weighing one pound. The reſt as in the two former harrows. The figure of each is annexed.

Theſe harrows I hold to be a conſiderable improvement. They ply to curved ground like two unconnected

unconnected harrows, and when drawn in one plain, they are in effect one harrow of double weight, which makes the teeth pierce deep into the ground. The imperfection of common harrows, mentioned above, will suggest the advantages of the set of harrows here recommended. The first is proper for harrowing land that has lain long after plowing, as where oats are sown on a winter-furrow; and in general, for harrowing stiff land: it pierces deep into the soil by its long teeth, and divides it minutely. The second is intended for covering the seed: its long teeth lays the seed deeper than the common harrow can do; which is no slight advantage. By placing the seed considerably under the surface, the young plants are protected from too much heat; and have sufficiency of moisture. At the same time, the seed is so well covered that none of it is lost. Seed slightly covered by the common harrows, wants moisture, and is burnt up by the sun; beside, that a proportion of it is left upon the surface uncovered. The third harrow supplies what may be deficient in the second, by smoothing the surface, and covering the seed more accurately. The three harrows make the ground finer and finer, as heckles do flax; or, to use a different comparison, the first harrow makes the bed, the second lays the seed in it, the third smooths the cloaths. These advantages are certain. If any man doubt, let him try the experiment, and he will

will find the effect of them in his crops. I can ſay ſo with aſſurance from the experience of many years. They have another advantage not inferior to any mentioned; they mix manure with the ſoil more intimately than can be done by common harrows; and upon ſuch intimate mixture depends greatly the effect of manure, as ſhall be explained afterward. To conclude, theſe harrows are contrived to anſwer an eſtabliſhed principle in agriculture, That fertility depends greatly on pulverizing the ſoil, and on an intimate mixture of manure with it, whether dung, lime, marl, or any other.

4. The ROLLER.

THE roller is an inſtrument of capital uſe in huſbandry, though ſcarcely known in ordinary practice; and, where introduced, it is commonly ſo ſlight as to have very little effect.

Rollers are of different kinds, ſtone, yetling, wood. Each of theſe has its advantages. I recommend the laſt, conſtructed in the following manner. Take the body of a tree, ſix feet ten inches long, the larger the better, made as near a perfect cylinder as poſſible. Surround this cylinder with three rows of fillies, one row in the middle, and one at each end. Line theſe fillies with planks of wood equally long with the roller, and ſo narrow as to ply into a circle. Bind them faſt

faſt together with iron rings. Beech wood is the beſt, being hard and tough. The roller thus mounted, ought to have a diameter of three feet ten inches. It has a double pair of ſhafts for two horſes abreaſt. Theſe are ſufficient in level ground: in ground not level, four horſes may be neceſſary. The roller without the ſhafts ought to weigh two hundred ſtone Dutch; and the large diameter makes this great weight eaſy to be drawn.

With reſpect to the ſeaſon for rolling. Rolling wheat in the month of April, is an important article in looſe ſoil; as the winter-rains, preſſing down the ſoil, leave many roots in the air. Barley ought to be rolled immediately after the ſeed is ſown; eſpecially where graſs-ſeeds are ſown with it. The beſt time for rolling a gravely ſoil, is as ſoon as the mould is ſo dry as to bear the roller without clinging to it. A clay ſoil ought neither to be tilled, harrowed, nor rolled, till the field be perfectly dry. And as rolling a clay ſoil is chiefly intended for ſmoothing the ſurface, a dry ſeaſon may be patiently waited for, even till the crop be three inches high. There is the greater reaſon for this precaution, becauſe much rain immediately after rolling is apt to cake the ſurface when drought follows. Oats in a light ſoil may be rolled immediately after the ſeed is ſown, unleſs the ground be ſo wet as to cling to the roller. In a clay ſoil, delay rolling till the grain

grain be above ground. The proper time for ſowing graſs-ſeeds in an oat-field, is when the grain is three inches high; and rolling ſhould immediately ſucceed whatever the ſoil be. Flax ought to be rolled immediately after ſowing. This ſhould never be neglected; for it makes the ſeed puſh equally, and prevents after-growth, the bad effect of which is viſible in every ſtep of the proceſs for dreſſing flax. The firſt year's crop of ſown graſſes ought to be rolled as early the next ſpring as the ground will bear the horſes. It fixes all the roots preciſely as in the caſe of wheat. Rolling the ſecond and third crops in looſe ſoil, is an uſeful work; though not ſo eſſential as rolling the firſt crop.

The effects of rolling properly uſed, are ſubſtantial. In the firſt place, it renders a looſe ſoil more compact and ſolid; which encourages the growth of plants, by making the earth clap cloſe to every part of every root. Nor need we be afraid of rendering the ſoil too compact; for no roller that can be drawn by two or four horſes will have that effect. In the next place, rolling keeps in the moiſture, and hinders drought to penetrate. This effect is of great moment. In a dry ſeaſon, it may make the difference of a good crop or no crop, eſpecially where the ſoil is light. In the third place, the rolling graſs-ſeeds, beſide the foregoing advantages, facilitates the mowing for hay. And it is to be hoped, that the advantage of this practice will lead farmers to mow their corn alſo, which will encreaſe the quantity of ſtraw, both for food and for the dunghill.

There is a ſmall roller for breaking clods in land intended for barley. The common way is, to break clods with a mallet, which requires many hands, and is a laborious work. This roller performs the work more effectually, and at much leſs expence: let a harrowing precede, which will break the clods a little; and after lying a day, or a day and half to dry, this roller will diſſolve them into powder. This however does not ſuperſede the uſe of the great roller after all the other articles are finiſhed, in order to make the ſoil compact, and to keep out the ſummer-drought. A ſtone roller four feet long, and fifteen inches diameter, drawn by one horſe, is ſufficient to break clods that are eaſily diſſolved by preſſure. The uſe of this roller in preparing ground for barley is gaining ground daily, even among ordinary tenants, who have become ſenſible both of the expence and toil of uſing wooden mells. But in a clay ſoil, the clods are ſometimes too firm, or too tough, to be ſubdued by ſo light a machine. In that caſe, a roller of the ſame ſize, but of a different conſtruction, is neceſſary. It ought to be ſurrounded with circles of iron, ſix inches aſunder, and ſeven inches deep; which will cut the moſt ſtubborn clods, and reduce them to powder. Let not this inſtrument be conſider-

ed

ed as a finical refinement. In a ſtiff clay, it may make the difference of a plentiful or ſcanty crop.

5. The FANNER.

THIS inſtrument for winnowing corn was introduced into Scotland not many years ago. Formerly wind being our only reſource, the winnowing of corn was no leſs precarious than the grinding it at a windmill: people often were reduced to famine in the midſt of plenty. There was another bad effect: it was neceſſary to place a barn open to the weſt wind, however irregular or inconvenient the ſituation might be with regard to the other buildings. But it is needleſs to be particular upon that uſeful inſtrument; becauſe every farmer conſiders it now as no leſs eſſential than a plough or a harrow.

CHAP. II.

FARM CATTLE and CARRIAGES.

1. FARM-HORSES.

A HORSE fit for a waggon, cart, or plough, ought to be ſtrong, compact, and about fifteen hands high. A carter or a ploughman cannot perform the ſame work with horſes of leſs ſize;

ſize; by which there is a conſiderable loſs, as he is paid by the year, not by the quantity of work he performs. Great attention ought to be given to the breaking a farm-horſe: good education will make him tractable and obedient to the voice, without neceſſity of applying the whip: the former makes the whole team move at once; the latter moves the horſe only that feels it.

We ſtop a little to conſider the diſadvantage of ſmall and ill-fed horſes, common in Scotland. Two ſtout horſes in a plough will make as deep a furrow as four of what are commonly uſed; and yet the former are leſs expenſive both in price and maintenance. A gentleman cannot do better for his own intereſt, than to promote a good breed of farm-horſes: two good horſes will be a ſaving of L. 8 Sterling yearly, that is expended by uſing four weak horſes. I ſhall mention only the carriage of lime. One ſervant fills his cart with a hundred ſtone, which two good horſes can pull with eaſe. Another lays but the half upon his cart, becauſe his two weak horſes are able for no more. This is a double loſs to the maſter: he gets leſs work, not from the horſes only, but alſo from the ſervant.

2. FARM-OXEN.

THERE is not in agriculture any other improvement that equals the uſing oxen inſtead of horſes; they

they are equally tractable; and they are purchased and maintained at much less expence. As this improvement is obvious to the meanest capacity, one might expect to see every farmer greedily embracing it, as he would a feast after being famished. Yet few stir. How is this to be accounted for? Men are led in chains by custom; and fettered even against their interest. "And "why should we pretend to be wiser than our "fathers?" they will say modestly, or rather obstinately.

What warms me upon this subject, is the great consumption of oats by work-horses, which would be totally saved by using oxen only. Did our own product furnish this consumption, it were less to be regretted; but it is grievous to be reduced to the necessity of importing annually vast quantities of oats; all of which would be saved by employing oxen only in a farm. Nor is this all that would be saved, as shall be mentioned by and by.

But that I may not be accused of declaiming without foundation, I am willing to enter into a candid comparison between horses and oxen as employ'd in a farm. I begin with affirming, that an ox is as tractable as a horse, and as easily trained to a plough or a cart. I have seen a couple of them in a plough going as sweetly without a driver as a couple of horses; directed by the voice alone without a rein. Oxen beside are preferable

preferable for a ſteady draught, as they always pull to their ſtrength, without ever flinching: horſes, on the contrary, are apt to ſtop when they meet with unexpected reſiſtance. As oxen have leſs air and ſpirit in moving than horſes, their motion is concluded to be ſlower. They are leſs expeditious, it is true, in galloping, or perhaps in trotting; but as farm-work is performed by ſtepping, let the ſtep of a horſe and of an ox be compared, and the latter will be found not inferior, eſpecially where an ox is harneſſed like a horſe. Colonel Pool in Derbyſhire plows as much ground with three oxen, as the neighbouring farmers do with four or five horſes. In ſummer they eat nothing but graſs: in winter, they have hay or turnip when much wrought; ſtraw only when wrought moderately. About Bawtry, in Yorkſhire, four oxen in a plough do as much work as the ſame number of good horſes. In ſeveral parts of Kent, an acre daily is plowed with a team of oxen, ſometimes a quarter more. Near Beaconsfield, Mr Burke plows an acre in a day with four oxen; and his neighbours do no more with four horſes. In the road from Leeds to Wetherby, I ſaw a loaded cart drawn by two ſtout horſes and a bull, all in a line, the bull in the middle. That draught was not ſlower than thoſe before and after in the ſame road. And ſurely the bull would not have been added had he retarded the horſes.

Hitherto

Hitherto the comparifon holds pretty equal. In one article oxen are clearly preferable. Their dung makes excellent manure; and by that means they always improve the pafture. Horfe-dung, on the contrary, burns where it falls, and hurts the pafture. Horfe-dung from the ftable has a greater tendency to burn than to rot; and to make it ufeful, it requires to be carefully mixed with cooler materials.

But the chief advantage of oxen comes under the article of favings, which branch out into many particulars. In the firft place, the price of a horfe fit for labour doubles that of an ox. An ox worth feven pounds, will perform as much folid work as a horfe worth fourteen. This is an important article: the labouring cattle are the moft expenfive part of a farm-ftock; and it is that expence which keeps back from farming many men whofe fkill and induftry would afford them a comfortable living. In that view, it is greatly the intereft of landlords to promote oxen, as they tend to multiply candidates for a farm; which not only gives the landlord opportunity for a proper choice, but raifes every farm to its juft value.

As an ox is cheaper than a horfe, fo he is fed cheaper in proportion. He requires no corn, and he works to perfection upon cut grafs in fummer, and upon hay in winter. He does well even upon oat-ftraw. Thus by ufing oxen, a farmer

mer can make money of his whole crop of oats, except what is neceſſary for maintenance of his family. The bulk of that product, on the contrary, is conſumed by farm-horſes. Even in the carſe of Gowry, the conſumption of oats on farm-horſes is ſo great, that at Perth and Dundee, there are annually imported between four and five thouſand bolls of oat-meal.

A horſe is liable to many diſeaſes that an ox is free from. If he happen to turn lame, to which he is ſubjected from many accidents, he is rendered uſeleſs. An ox may always be turned to account; for if diſabled from work, he can be fatted for the ſhambles, and ſold for more than was paid for him.

A horſe commonly turns uſeleſs for work in ten years, and the ſtock of horſes muſt be renewed every ten years at a medium, which is a deep article of expence to the farmer. Oxen laſt for ever; or, which comes to the ſame, they can be ſold to the butcher when paſt the vigour of work, and their price will be more than ſufficient to put young oxen in their ſtead.

Horſes require more attendance than oxen: they muſt be curried, combed, and rubbed down. Let oxen have their proper quantity of food, and they require no other care. It is ſufficient employment for a man to manage four or five horſes: he will manage with equal eaſe double the number of oxen.

The

The ſhoeing of horſes is no inconſiderable article. The expence of ſhoeing oxen is a mere trifle.

Theſe ſeveral articles of ſaving are ſummed up in a following table, and are very conſiderable. This ſum ought to go wholly to the landlord as additional rent. The tenant has no claim for any ſhare; becauſe after paying that additional rent, he has as much profit as he had formerly when he wrought with horſes.

By this mode of huſbandry, the advantage to the landlord is great; and to the kingdom much greater, by ſaving the importation of an immenſe quantity of oats. But the advantage of oxen is ſtill more extenſive: it reaches every manufacturer, and indeed the whole people. There muſt be a great increaſe of oxen to anſwer the purpoſes of farming: every one of theſe, after their prime is over, goes to the ſhambles: the markets are filled with beef, which not only lowers the price of beef, but of leather and tallow. The ſavings upon theſe articles would bring down the wages of our manufacturers, and conſequently the price of our manufactures in a foreign market; not to mention that cheap manufactures at home tend alſo to lower wages.

People differ in the manner of yoking oxen. In ſome places they are yoked by the tip of the horn; in ſome by the root. Theſe modes are viſibly inconvenient. When an ox draws by the ſhoulder like a horſe, his head is free, and his motion natural. When yoked by the horns, he lowers his head to the line of the draught: his poſture is conſtrained, and his ſtep ſhort. His neck indeed is ſtrong, but his ſhoulder is a better *fulcrum* for the draught. To yoke an ox by the ſhoulder, his harneſs ought to be the ſame with that of a horſe. The only difference is, that as his horns hinder the collar from being ſlipped over his head, it muſt be open below, and buckled after it is on. The advantage of yoking an ox by the ſhoulder was known even in the time of Columella; who ſays that faſtening the yoke to their horns, is rejected by almoſt all who have written directions for huſbandmen; for the cattle can exert greater efforts with the breaſt than with the horns. Book 2. chap. 2.

When the advantages of oxen for draught are ſo great, it cannot but appear ſtrange, that in Britain oxen have almoſt totally been laid aſide. Among the ancients, we read of no beaſts for draught but oxen. It was ſo in Greece, as early as the days of Heſiod; and it was ſo every where. The Dutch at the Cape of Good Hope plow with oxen, and exerciſe them early to a quick pace, ſo as to equal horſes in the waggon, as well as in the plough. They are uſed in the Eaſt Indies for carrying burdens; and they are fitter even than horſes for that ſervice, the back of an ox being convex, and conſequently more able to

ſupport

ſupport a weight than that of a horſe. All that is neceſſary in the furniture for the back, is a bit of wood or ſtiff leather, to prevent the load from falling down upon the neck. The only cauſe I can aſſign for preferring horſes, are bad roads, which were univerſal in Britain till lately. Being impracticable for carts during winter, the farmer carried his corn to market on horſeback. A proper furniture for the back of an ox was not thought of, though an eaſy invention. And horſes being thought neceſſary for carrying burdens, they were employed inſtead of oxen in every work: if employed at all, they are too expenſive ever to be ſuffered to be idle. Another circumſtance contributed. Becauſe oxen require no corn, it is commonly imagined that they ſcarce require any food. They are put off during winter with dry ſtraw, which after the turn of the year affords very little nouriſhment. They become too weak for working; and yet, inſtead of bettering their food, it is vainly thought that multiplying their number will anſwer; and thus may be ſeen in ſeveral places yoked in a plough, ten or twelve weak animals that can ſcarce ſupport their own weight. We are now provided with good roads every where; and there is no longer the pretext of bad roads for preferring horſes. Corn is now carried to market in carts, for which oxen are no leſs proper than horſes. And it is hoped, that farmers will at laſt break through

through a bad cuſtom, and open their eyes to their own intereſt. Nothing is more deeply their intereſt than to lay aſide horſes totally in farm-operations, and to employ oxen. The tackſman profits firſt; but does not the landlord gain more, by enabling his tenants in new leaſes to pay a higher rent? Why then ſhould gentlemen loitre, while they can ſo eaſily advance their rent without oppreſſing their tenants? Why do they not encourage their tenants by example and precept, to follow a mode that is equally beneficial to themſelves and to their country? It will be hard indeed, if a ſingle tenant cannot be found to ſee his intereſt: if a landlord can prevail but upon one or two of his tenants to take the lead, the reſt will naturally follow. At any rate, he can force them to their own good, by prohibiting horſes in every new leaſe. It is a ſtrange ſort of ambition that moves gentlemen to ſpend their eſtates in the Houſe of Commons, where moſt of them are mere mutes, inſtead of ſerving their country and themſelves at home, which is genuine patriotiſm *.

As

* Columella, Book 2. chap. 2, adviſes the ploughman to give his oxen a little reſt at the end of every ridge. "But, ſays he, a longer ridge than one hundred and "twenty feet is hurtful to the cattle, by fatiguing them "more than they ought to be." Oxen are more fatigued with heat than horſes; which appears even in this cold country

As computation is the touchſtone of profit and loſs, two computations are ſubjoined; one to ſatisfy the farmer of the ſum he will ſave by employing oxen inſtead of horſes; and one to ſhew what benefit will accrue to the public by the change. To ſet the firſt computation in a clear light, and to avoid fractions, I make the ſuppoſition, that a horſe put to work at the age of five may endure hard work for twelve years, which is a large allowance beyond the truth. An ox is put to work at the age of four, and at ſeven is in his prime, which is the proper time to feed for the ſhambles. The computation accordingly is framed upon a revolution of twelve years; during which period oxen are four times changed without any change of horſes. At the end of the period, both muſt be changed; and a new revolution goes on as before.

TABLE

country during the heat of ſummer. Yet in the hotteſt countries oxen are preferred for labour; how much more in a cold country like Scotland. A yoke of oxen among the old Romans commonly plowed a *jugerum* in a day, which is nearly equal to two-thirds of an Engliſh acre; two Engliſh acres making about three *jugera*. Our Saxon anceſtors had their *bovata teræ* or "ox gang," which was fifteen acres; ſix of which made a plough-land, *viz.* as much as ſix oxen can plow in a year.

TABLE, shewing how much a farmer saves by employing Oxen instead of Horses.

	L.	s.	d.	L.	s.	d.
A Horse of 5 years old is purchased or valued at				15	0	0
To grasing ditto 24 weeks, at 2 s. 1 d. per week	2	10	0			
To corn in winter 15 weeks, 2 pecks per week, at 8 d. per peck -	1	0	0			
To corn in spring 13 weeks, 4 pecks per week, at 10 d. per peck -	2	3	4			
To shoeing and farrying one year - -	0	12	6			
To insurance against lameness and death -	0	7	6			
One year's maintenance	6	13	4			
Twelve years maintenance				80	0	0
Total amount of the first cost, and 12 years maintenance - -				95	0	0
The horse unfit for hard labour may be sold for - -				3	0	0
The expence of one horse at the end of twelve years - -				92	0	0

The

The expence of an Ox twelve years.

	L.	s.	d.	L.	s.	d.
An Ox * 4 years old purchafed at - -				5	10	0
To grafing 24 weeks at 1 s. 6 d. per week -	1	16	0			
To fhoeing and farrying one year - -	0	4	6			
To infurance againft accidental death -	0	2	6			
Amount of one year's maintenance -	2	3	0			
Maintenance for 3 years				6	9	0
Coft and maintenance -				11	19	0
Deduce the price he can be fold at				7	10	0
Total coft at the end of 3 years -				4	9	0
Total coft at the end of 12 years -				17	16	0
Subtract this fum from the expence of a horfe during the fame time,				92	0	0
The balance againft the horfe is				74	4	0

By

* An ox light made and of a middle fize, is of all the fitteft for the plough ; and, in many places of Scotland, fuch

By this table it appears, that the yearly expence of a farm-horse exceeds that of an ox in the sum of L. 6 : 3 : 8. Now supposing four horses necessary to a farm of 100 acres, rent L. 50 yearly, four good oxen well fed will perform the same work; and therefore the using the latter instead of the former, will be an annual saving of L. 24 : 14 : 8, very near the half of the rent. And what will raise it above half of the rent, is the interest of the money laid out, which is considerably higher upon the horse than upon the ox; but which is omitted in the foregoing table, to avoid an intricate calculation.

There is beside an article that preponderates greatly for the ox. His dung enriches the ground on which he pastures. Horse-dung, on the contrary, is hurtful. This is an article of great importance.

N. B. The hay or straw consumed by the horse in winter is not less than that consumed by the ox; and therefore that article is left out with respect to both.

Computation

such an ox may be purchased under six pounds. But an ox reared in a rich soil and that gets plenty of food, will at the same age draw a much higher price; and, accordingly, computing the profit of rearing horned cattle, I have stated the value at seven pounds.

Computation of the yearly quantity of oats consumed in Scotland by labouring horses.

FOR repairing the highways in East Lothian, each farm that is managed with a single plough is rated at a certain sum; and there are in the county computed to be 1331 such farms. In every plough four horses are employed; and many farmers employ additional horses for harrowing, *&c.*; but supposing only four, the number of farm-horses will be 5324. According to the preceding table, the yearly quantity of corn consumed by a horse is 5 bolls 2 pecks; consequently the amount of the corn consumed annually is 27,285½ bolls; which, computed at 12 s. a-boll, a moderate price, amounts to the sum of L. 16,371, 6 s. Sterling.

Reckoning the horses employed in all the other corn-countries in Scotland, their number will be at least fifteen times as many as what are employed in East Lothian. Therefore, to find out the value of the oats consumed yearly in Scotland by work-horses, multiply the said sum of L. 16,371, 6s. Sterling by 16, which is L. 261,940, 16 s.; an immense sum that would be saved to this nation by employing oxen in agriculture instead of horses. In one of Young's tours there are computed to be in England 684,491 draught cattle, of which the horses must amount at least to

 680,000;

680,000; each of which may be reckoned to consume six bolls of oats yearly. And reckoning a boll of oats at twelve shillings, the value of oats consumed will amount to L. 2,448,000; all of which may be saved by employing oxen.

If I cannot prevail upon farmers to mind their own interest by employing oxen instead of horses, I shall at least make them blush, by comparing them with farmers no less obstinate in another particular. A farmer was brought from Northfolk to the neighbourhood of Axminster, who being disgusted at the husbandry practised there, commenced a better plan. His first attempt was, to make turnip a regular crop in rotation, and to hoe them twice in the Norfolk manner. He met with many obstructions from perverseness and aukwardness in his servants; but by perseverance, and working with them himself, he has prevailed. He now for many years has had exceeding good crops; which at the same time have improved the succeeding crops of barley. Of these facts his neighbours have been witnesses near twenty years; and yet not one of them has followed his example. Can there be a stronger instance of prejudice, or rather stupidity?

3. Breeding Horses and Horned Cattle.

In good husbandry, the greatest profit that can be made of grass is the breeding horses for sale. The

The profit will appear from the following table.

The ſtallion,	L.	0	15	0
Summer-graſs for the mare and foal,		2	0	0
Expence of the foal the firſt winter,		1	15	0
Summer-graſs ſecond year, - -		1	5	0
Winter-graſs, - - - -		0	15	0
Summer-graſs thind year, -		1	10	0
Winter-graſs, - - - -		0	15	0
Summer-graſs fourth year, -		1	15	0
Winter-graſs with corn, - -		1	10	0
Total expence,	L.	12	0	0
Being now riſing five years old, his value, - - -		16	0	0
Clear profit,	L.	4	0	0

A young horſe after the firſt winter need not be houſed till the fourth winter, provided he have a ſhed to fly into in a ſtorm, with a little hay or ſtraw.

It is ſtill more profitable to buy young horſes three years old, and at five to diſpoſe of them, or work them in the farm.

A horſe is a great eater. In common outfield-graſs it will take two acres for his ſummer-paſture. And even where ground has been laid out with graſs-ſeeds, it will take an acre and a half after the ſixth or ſeventh year.

The breeding of horned cattle is alſo profitable, and

and has a peculiar advantage, namely, that a beaſt bred on a farm, thrives always beſt there. The profit appears from the following table.

	L.	s.	d.
For the bull, - - -	L. 0	2	6
For milk during ſummer, - -	1	0	0
For wintering, - - -	0	10	0
Summer-graſs ſecond year, - -	0	16	0
Winter-food, - - -	0	10	0
Summer-graſs third year, - -	1	4	0
Winter-food, - - - -	0	15	0
Total expence,	L. 4	17	6
A young bullock entering his fourth year, will ſell for - -	7	0	0
Clear profit,	L. 2	2	6

4. Wheel-Carriages.

Wheel-Carriages for a farm are, waggons drawn by four horſes; carts by one, two, or three horſes; carts drawn by oxen in yokes, termed in Scotland *coups;* and a cart with three wheels by one horſe. Till lately our farmers had no wheel-carriages; and to this day they are not univerſal.

Whatever ſtrength a horſe may have, yet the weight he can draw is determined by his own weight:

weight: in drawing, there is a certain weight that he cannot exceed without being raiſed off the ground; and therefore, to enable him to exert his whole ſtrength, ſome weight ought to be laid on his back. Nothing is more common than to ſee a carter mount the ſhaft-horſe, when hard ſtrained in drawing up-hill. He imagines that he has the horſe more under command; but the true reaſon is, that the horſe draws more by having weight on his back. In the ordinary way of yoking two horſes in a line, the ſhaft horſe, who is burdened with part of the weight of the cart, draws much more than the other horſe can do.

I have taken ſome pains to know what can be drawn in a cart without ſtraining the horſes; but I find no uniformity. Two horſes in a cart, yoked in a line, uſually draw, from Borrowſtounneſs to Glaſgow, three thouſand weight. From Stirling to Glaſgow they draw but twenty-four hundred weight. Alexander Monteith, a carter in the neighbourhood of Carron, has, with two horſes, repeatedly drawn from Banton, in the pariſh of Kilſyth, to Carron, four thouſand four hundred weight, each hundred weight conſiſting of one hundred and twenty pounds Avoirdupois. He commonly draws thirty-ſeven or thirty-eight hundred weight. The carts between Borrowſtounneſs and Glaſgow carry from ten to fifteen hundred weight with a ſingle horſe. And with the ſame

ſame draught they carry about Edinburgh twelve hundred weight of coals. They carried formerly no more but that weight with two horſes yoked in a line. Mr Orr of Barrowfield, for carrying his coals to market, uſes ſingle-horſe carts with wheels ſix feet high. Every cart carries eaſily twenty-one hundred weight to the place of unloading, which is diſtant from the coal-pit about two Engliſh miles; and this is done ſix times every day. The inequality of theſe weights ſhows the inaccuracy of carters, who have not come to any preciſe knowledge of what horſes can perform in a cart. In the mean time, till more exact experiments be made, we ſhall take a middle rate, which is the three thouſand weight drawn by two horſes from Borrowſtounneſs to Glaſgow, the half of which for a ſingle horſe is fifteen hundred weight; though, conſidering the weight that lies upon the back of a ſingle horſe in a cart, his part of the draught may be well computed at ſixteen hundred weight.

This leads to a compariſon between a waggon and ſingle-horſe carts of the ordinary make. To keep within bounds, we may fairly take it for granted, that in a well conſtructed cart, a ſingle horſe of moderate ſize will command fourteen hundred weight; conſequently that ſix horſes in ſix carts will draw eight thouſand four hundred weight. Let us now ſee what is drawn in a waggon with ſix horſes. The ordinary weight in this

this country are four tons, or eight thousand weight; and I am informed it is the same about London. At that rate, six horses in six ordinary carts, draw four hundred weight more than six horses in a waggon. I suppose that the weight of the six carts will be at least equal to that of a waggon. At the same time, in a turnpike-road, a man and a boy are sufficient to manage the former as well as the latter.

But, to carry on the comparison, small carts have another advantage, which is, that they admit high wheels; and it is easy to adjust the wheels to the height of the horse, by making the axle go through the middle of the cart, or higher if necessary, to make the horse draw horizontally. It is a great ease to the horse to make the axle go through the centre of gravity; for by that means the weight on the horse's back is the same going up-hill or down-hill. A waggon carrying four or five tons, is incapable of that improvement; for high wheels able to support such a weight, would require a strength of timber that would be intolerable, especially in going up-hill. Here appears in a conspicuous light the advantage of single carts with high wheels above a waggon. Upon Mr Orr's practice, I take it for granted, that with wheels six feet high, a single horse will carry easily twenty-one hundred weight for a whole season. But let us state only two thousand weight, or a single ton, is not the conclusion fair, that in such

carts

carts ſix horſes will draw a third more than in a waggon?

But to be more particular, two loaded carts with wheels as high as the hind wheels of a waggon will be eaſier drawn by three horſes in each, than the ſame load in a waggon with ſix horſes. The advantage of the waggon is, that the ſhaft-horſe has nothing to bear. On the other hand, the lateral ſhake from rough ground, ſo ſevere in the waggon, is divided between two horſes in two carts. Here both have their advantages and diſadvantages. To give the advantage of the waggon without the lateral ſhake, let ſix ſingle-horſe carts be uſed inſtead of a waggon, with the body below the axle and hung upon it ſo as to play freely. The diſadvantage of twelve wheels inſtead of four, will be more than balanced by relieving the horſes from the oppreſſion of weight and from the lateral ſhake; and their ſtrength will be intirely applied to pulling. At any rate, ſix ſingle carts will draw more than is allowed to a waggon by the late act of parliament; which is no more but four ton beſide the weight of the waggon.

Thus, ſingle-horſe carts are clearly preferable to a waggon with reſpect to the burden they can carry; and they are ſtill more preferable with reſpect to the highways. A turnpike-road would be eaſily made, and more eaſily ſupported, were none but ſingle-horſe carts admitted, or were the toll

toll doubled or tripled upon waggons. This ſpeculation merits the moſt ſerious attention of the legiſlature. To preſerve turnpike-roads in perfect order, a ſingle act of parliament would be ſufficient. The Iriſh have taken the lead in this important article. We ought to follow without a moment's delay.

The proportion that ought to be laid on a horſe's ſhoulders to have the greateſt command of the draught, remains to be aſcertained by experiments. But I gueſs, that a horſe who can draw fourteen hundred weight without any load on his back, will with equal eaſe draw ſixteen hundred, by laying on his back five or ſix ſtone of that weight.

A wheel-carriage is a great ſaving to the farmer, much greater than in a curſory view will be imagined. A cart with two horſes carries five bolls of ſhell-lime wheat-meaſure. I talk of ordinary carts with two horſes in a line; for two horſes abreaſt in ſhafts will carry much more. Six horſes commonly are uſed for carrying the ſame quantity on their backs. To the former one man is ſufficient; to the latter, one to every two horſes. Now, ſuppoſing a man and a cart with two horſes, to be hired for three ſhillings a-day, the expence of the lime on horſeback is nine ſhillings; in a cart, no more but three. Here is a ſaving of ſix ſhillings upon every five bolls. And ſuppoſing forty-eight cart-loads of lime to be led in a ſummer, which is a moderate computation even in a farm that employs but a ſingle plough, the ſaving amounts to L. 14, 8 s. which alone is ſufficient to convert a loſing farm into a profitable one. I have been a witneſs to the carrying on horſeback 700 loads of coals, a man to every two horſes; the expence of which amounts to L. 52, 10s. ſuppoſing a man to be hired at eight pence a-day, and a horſe at fourteen pence. That quantity can be carried in carts with two horſes for L. 21 : 2 : 11. The ſame calculation is applicable to grain. A cart with two horſes will carry to market ſix bolls of barley, which, when carried on horſeback, require ſix horſes. A ſingle plough in tolerable ſoil well cultivated, will afford for ſale a hundred bolls of grain, which carried in carts makes a ſaving of L. 4. Sterling.

The uſe of a three-wheeled cart drawn by a ſingle horſe, is to remove earth, or to carry manure to a near field. One horſe ſerves two of theſe carts: when he returns after unloading, the other is ready filled. But this work is ſo ſevere for a ſingle horſe, that the field ought to be ſmooth and level; or inclining downward, which is ſtill better.

I ſhall only add upon this head, that the ſacks commonly uſed in Scotland for corn, are too large. In England ſmall ſacks are uſed, which one man can eaſily load or unload.

CHAP.

CHAP. III.

Farm-Offices.

IN this country there are few traces of ſkill or contrivance in farm-houſes; no regard to a centric ſituation, nor to a dry ſpot, nor to ventilation. Our farm-offices are ſet down ſtraggling and confuſed as if by accident; here a barn, there a ſtable. Imperfection in form is more excuſable, there being few good examples to copy from: every where cow-houſes ſo aukwardly formed, that they cannot be cleaned till the cattle be turned to the door; and ſo ſtrait that they muſt dung upon each other. And laſtly, after all the labours of the year, no place contrived for keeping corns dry.

Though to a ſtack-yard dryneſs of ſituation and free ventilation are eſſential; yet ſo little are theſe circumſtances regarded, that it is always adjacent to the dwelling-houſe, whether the ſpot be wet or dry. It is at the ſame time carefully ſurrounded with trees, as if to prevent ventilation, and as if water dropping from the branches on the ſtacks were ſalutary to them.

A kitchen-garden is of importance to a farm, as will appear afterward. There is indeed always the name of a kitchen-garden, but very little of the reality. The chief attention is to ſurround

ſurround it with trees; and yet the neceſſary effect of excluding free air, is to dwarf the plants, and to give them a bad taſte.

It ſeems to be the opinion of our farmers, that a dunghill cannot be too moiſt; for it is commonly put in a hole, and conſequently ſurrounded with water: the richeſt parts are imbibed by the water, and both evaporate together, leaving the dunghill little better than a *caput mortuum*. Water at the ſame time, above a very moderate proportion, is far from contributing to putrefaction. I have ſeen a ſheaf of ſtraw, after lying ſix months in water, ſo tough as to be fit for making ropes.

The foregoing defects are but imperfectly remedied in the lateſt conſtructions. The form moſt in requeſt is three ſides of a ſquare; the houſes for the farm-cattle on the eaſt, the barn on the weſt, and ſheds on the north, the dunghill occupying the middle of the ſquare. The ſtables are too far from the barn; the dung lies ſcattered, and trodden upon by the cattle; and the expence of roofing is great.

Theſe particulars are mentioned as an introduction to the following plan; preferable upon two accounts; firſt, as leſs expenſive; and next, as more convenient. The plan is, to erect a houſe of two ſtories, the under ſtory for a ſtable, and the upper for a barn. The door of the ſtable fronts the eaſt; that of the barn the weſt, having a ſtair that leads up to it. The ſtack-yard joins the barn, with free air to the north, weſt, and ſouth. Round this building are ſheds for cattle laid to the walls, with roofs ſloping from the eaſing. The ſheds ſhould be twelve feet wide, ſufficient for the ſize of any farm-beaſt; and the outward wall may be ſeven feet high or ſo. There is place for ſix ſheds, one on each gable, and two on each ſide-wall, leaving an interval for the doors of the barn and ſtable.

There will be found a great ſaving in this plan, compared with the other mentioned. To form a juſt compariſon, they ought to be equally capacious; the two barns ought to be of the ſame ſize, and there ought to be room for the ſame number of cattle. The expence of each is annexed; and the ſaving on the plan recommended is no leſs than L. 94 : 2 : 2 Sterling. This is an important article to every gentleman who poſſeſſes an eſtate in a corn-country; for no article is heavier than the building farm-houſes. Next as to convenience. In the ordinary form, where the barn is on the one ſide of a court, and the ſtables on the oppoſite ſide, the time conſumed in carrying ſtraw to the cattle is almoſt entirely ſaved in the plan propoſed; there being holes in the barn-floor over every ſtall, covered with a moveable board, through which ſtraw is let down to the cattle; which abridges greatly the labour of the ploughman and carter. At the ſame

ſame time, to leave ſuch work to ſervants without any check, is often made a ſcreen for idleneſs. With reſpect to threſhing, a timber-floor has great elaſticity; and I am made certain from experience, that a third part more may be threſhed on it than on an earthen floor, which has no elaſticity. Further, the frequent ſweeping of an earthen-floor raiſes the finer parts of the earth, which mix with the grain: this is totally prevented in cleaning corn upon a timber-floor. Add that the dampneſs of an earthen floor corrupts the grain; it cannot be kept a fortnight from vegetating, eſpecially if laid up to the damp wall. Sacks give no ſecurity; for they rot if allowed to ſtand in the barn any time. A barn in a ſecond floor is excellent for preventing all theſe evils: one end contains a ſtack of corn, the middle is for threſhing, and the other end for cleaning the corn when threſhed. Above this end, there is a loft for holding the cleaned corn, to which there is eaſy acceſs by a ſtair; and here corn may be kept in ſafety for months.

It will be objected againſt this plan, that there is no court for a dunghill, where winterers are commonly fed. But this I have purpoſely avoided; for it will be ſeen afterward, in the inſtructions given for the feeding of cattle, that to keep winterers in that manner is hurtful both to them and to the dunghill. They will be more comfortably put up in ſheds; provided only in building

ing care be taken to give a free circulation of air.

One article ſtill remains, in treating of farm-offices, which is a houſe for laying up inſtruments of huſbandry, when not employed. Few farmers are ſufficiently careful about this article: they behave as if inſtruments of huſbandry could be procured without pains, and without price. It is true indeed, that theſe inſtruments are generally ſo mean and ſo inſufficient as to deſerve very little care; — nothing more common, than to be left where laſt uſed, open to heat and cold, drought and moiſture.

To form a juſt notion of the propereſt method for preſerving wood, I muſt premiſe, that it will laſt equally well in air and in water; but it muſt be kept conſtantly in the one or the other. What deſtroys wood is the alternate action of air and water. Obſerve a ſtake driven into the ground: the part that conſumes firſt, is not that under the ſurface, nor that freely above it, but the ring at the ſurface, to which air and moiſture have equal acceſs. The ſame, where one piece of wood is mortiſed into another: the part that decays firſt, is where the two pieces join, which is open to the air, and at the ſame time lodges moiſture. The ſame is obſervable in the putrefaction of a dunghill: the parts from which either air or water is excluded, never rot. Though in a mortiſe the parts joined rot the ſooneſt, yet the mortiſed part, from which both air and water are excluded

cluded, decays ſooner than that which is open to the air, ſuppoſing it to be kept dry.

Hence it follows, that to preſerve wood in the moſt perfect manner, it ought to be ſheltered from rain, and expoſed as much as poſſible to the ventilation of dry air. This ſuggeſts the beſt conſtruction of a houſe for preſerving inſtruments of huſbandry. It ought to be erected in the higheſt ſpot, free to every wind: it ought to have a roof ſupported on pillars; the ſides ought to be conſtructed like thoſe of a drying-houſe at a bleach-field, with moveable boards for admitting air and excluding rain. Three feet or ſo may be left open at bottom; becauſe ſo low down no rain can penetrate to do miſchief. Need I add, that before laying up any inſtrument, it ought to be carefully cleaned and dried?

Expence of Farm-offices round a Farm-yard.

Building 18 roods 2 yards, at L. 1, 5 s. per rood, - - -	L. 23	6	4
Logs, 1120 feet, at 12 d. per foot,	56	0	0
Deals, 1148, at 12 d. per yard,	57	8	0
Binding the roof and ſarking, at 2 d. per yard, - - - -	9	11	4
Slates, 52,000, at 16 s. per 1000, -	41	12	0
Nails, 86,000, at 5 s. per 1000, -	21	10	0
Slating 31 roods 32 yards, at 16 s. 8 d. per rood, - - - - -	26	0	10
8 doors, at 7 s. each, - -	2	16	0
	L. 238	4	6

Ex-

Expence of the form recommended.

		L.	s.	d.
Building 13 roods 6 yards, at L. 1, 5s. per rood,	- - -	L. 16	8	0
Logs, 615 feet, at 12 d. per foot,		30	15	0
Deals for sarking, 576 yards, at 12 d. per yard,	- - -	28	16	0
Binding the roof, 2 d. per yard,		4	16	0
Slates, 32,400, at 16 s. per 1000,	-	25	18	8
Nails, 43,000, at 5 s. per 1000	-	10	15	0
Slating 19 roods 4 yards, at 16 s. 8 d. per rood,	- - -	16	13	0
Deals for flooring, 85, at 16 d. each,		6	1	8
Laying the floor, 88 yards, at 6 d. per yard,	- - - -	2	4	0
Doors 5, at 7 s. each,	- -	1	15	0
		L. 144	2	4

N. B. The expence of the house for holding the husbandry instruments, is not taken into this account; because every farm ought to be provided with such a house, whatever be the construction of the other farm-offices.

CHAP. IV.

Preparing Land for Cropping.

1. Obstructions to Cropping.

IN preparing land for cropping, the firſt thing that occurs to the writer and to the huſbandman, is to conſider the obſtructions to regular plowing. The moſt formidable of theſe, are ſtones lying above or under the ſurface, which are an impediment to a plough, as rocks are to a ſhip. Did not cuſtom account for it, how ſtrange would it appear, that few proprietors or tenants in Scotland think of clearing their land of ſtones. Stones above the ſurface may be avoided by the ploughman, though not without loſs of ground; but ſtones under the ſurface are commonly not diſcovered till the plough be ſhattered to pieces, and perhaps a day's work loſt. The clearing land of ſtones is therefore neceſſary to prevent miſchief. And to encourage the operation, it is attended with much actual profit. Take the following particulars. The ſtones are uſeful for fences: when large they muſt be blown, and commonly fall into parts proper for building. And as the blowing, when gunpowder is furniſhed, does not exceed a halfpenny for each inch that is bored, theſe ſtones come cheaper than to dig

dig as many out of the quarry. In the next place, as the ſoil round a large ſtone is commonly the beſt in the field, it is purchaſed at a low rate by taking out the ſtone. Nor is this a trifle; for not only is the ground loſt that is occupied by a large ſtone, but alſo a conſiderable ſpace round it, to which the plough has not acceſs without danger. A third advantage is greater than all the reſt; which is, that the plowing can be carried on with much expedition, when there is no apprehenſion of ſtones: in ſtony land, the plough muſt proceed ſo ſlowly, as not to perform half of its work.

To clear land of ſtones, is in many inſtances an undertaking too expenſive for a tenant who has not a very long leaſe. As it is profitable both to him and to his landlord, it appears reaſonable that the work ſhould be divided, where the leaſe exceeds not nineteen years. It falls naturally upon the landlord to be at the expence of blowing the ſtones, and upon the tenant to carry them off the field.

It is vain to think of drawing any conſiderable rent, till a farm be cleared of ſtones. Why then do gentlemen neglect this means of improving their land? In a leaſe, let it be a proviſo, that the landlord or his ſteward be advertiſed of every ſtone that obſtructs the plough. When a number of theſe are marked, let an artiſt be employed to bore and to blow; and the landlord has done

done his part. I engage that he will make twenty *per cent.* of the money laid out in this operation.

Another obſtruction is wet ground. Water is a good ſervant, but a bad maſter. It may improve gravelly or ſandy ſoils; but it ſours a clay ſoil, and converts low ground into a moraſs, unfit for any purpoſe that can intereſt the huſbandman.

A great deal has been written upon different methods of draining land, moſtly ſo expenſive as to be ſcarce fit for the landlord, not to mention the tenant.

One way of draining without expence when land is to be encloſed with hedge and ditch, is to direct the ditches ſo as to carry off the water. But this method is not always practicable, even where the diviſions lie convenient for it. If the run of water be conſiderable, it will deſtroy the ditches, and lay open the fences, eſpecially where the ſoil is looſe or ſandy.

If ditches will not anſwer, hollow drains are ſometimes made, and ſometimes open drains, which muſt be made ſo deep as to command the water. The former is filled up with looſe ſtones, with bruſh-wood, or with any other porous matter that permits the water to paſs. The latter is left open and not filled up. To make the former effectual, the ground muſt have ſuch a ſlope as to give the water a briſk courſe. To execute them in level ground is a groſs error: the paſſages are ſoon

ſoon ſtopped up with ſand and ſediment, and the work rendered uſeleſs. This inconvenience takes not place in open drains: but they are ſubject to other inconveniencies: They are always filling up, to make a yearly reparation neceſſary; and they obſtruct both plowing and paſturing.

I venture to recommend the following drain as the beſt in all views. It is an open drain made with the plough, cleaving the ſpace intended for the drain over and over, till the furrow be made of a ſufficient depth for carrying off the water. The ſlope on either ſide may by repeated plowings be made ſo gentle as to give no obſtruction either to the plough or to the harrow. There is no occaſion for a ſpade, unleſs to ſmooth the ſides of the drain, and to remove accidental obſtructions in the bottom. The advantages of this drain are manifold. It is executed at much leſs expence than either of the former; and it is perpetual, as it never can be obſtructed. In level ground, it is true, graſs may grow at the bottom of the drain; but to clear off the graſs once in four or five years, will reſtore it to its original perfection. A hollow drain may be proper between the ſpring-head and the main drain, where the diſtance is not great; but in every other caſe the drain recommended is the beſt.

Where a level field is infeſted with water from higher ground, the water ought to be intercepted by a ditch carried along the foot of the high

high ground, and terminating in ſome capital drain.

The only way to carry off water from a field that is hollow in the middle, is a drain ſtill lower. This is commonly the caſe of a moraſs fed with water from higher ground, and kept on the ſurface by a clay bottom.

A clay ſoil of any thickneſs is never peſtered with ſprings; but it is peſtered with rain, which ſettles on the ſurface as in a cup. The only remedy is high narrow ridges, well rounded. And to clear the furrows, the furrow of the foot-ridge ought to be conſiderably lower, in order to carry off the water cleverly. It cannot be made too low, as nothing hurts clay ſoil more than ſtagnating water; witneſs the hollows at the ends of crooked ridges, which are abſolutely barren. Some gravelly ſoils have a clay bottom, which is a ſubſtantial benefit to a field when in graſs, as it retains moiſture. But when in tillage, ridges are neceſſary to prevent rain from ſettling at the bottom; and this is the only caſe where a gravelly ſoil ought to be ridged.

Clay ſoils that have little or no level, have ſometimes a gravelly bottom. For diſcharging the water, the only method I can think of, is, at the end of every ridge to pierce down to the gravel, which will abſorb the water. But if the furrow of the foot-ridge be low enough to receive all the water, it will be more expeditious to make a

few

few holes in that furrow. In ſome caſes, a field may be drained, by filling up the hollows with earth taken from higher ground. But as this method is expenſive, it will only be taken where no other method anſwers. Where a field happens to be partly wet, partly dry, there ought to be a ſeparation by a middle ridge, if it can be done conveniently. And the dry part may be plowed, while the other is drying.

The low part of Berwickſhire is generally a brick clay, extremely wet and poachy during winter. This in a good meaſure may be prevented by proper incloſing, as there is not a field but can be drained into lower ground, all the way down to the river Tweed. But as this would leſſen the quantity of rain in a dry climate, ſuch as is all the eaſt ſide of Britain, it may admit of ſome doubt whether the remedy would not be as bad as the diſeaſe.

Broom and whins are great obſtructions to cropping. Broom is an evergreen ſhrub that thrives beſt in ſandy ſoil; and there it grows ſo vigorouſly, as ſcarce to admit any graſs under it. A plant of broom that has arrived to its full ſize, dies when cut over: but this does not root out broom, becauſe it grows from ſeed lodged in the ground; beſide, that there can be no cropping while the ſtumps remain. An effectual way to root out this plant, is, after cutting the great ſtems cloſe to the ground, to carry them off the field.

And

And the ground may be cleared of the roots, by a Scotch plough with a ſpiked ſock, drawn by two oxen and two horſes.

The field thus cleared may bear a crop of oats, or two, and with proper manure may be continued in tillage, or laid down with the ſeed of paſture-graſſes. The latter is preferable from the nature of the ſoil, which is commonly ſandy.

But as the ſeed of broom lies long in ſuch ſoil without rotting, the farmer may reckon upon a plentiful crop from the ſeed along with the graſs. To pull up the young broom with the hand, is an expenſive work; and for a large field it is difficult to procure hands. Another method practiſed is, to cut the young broom with a ſythe. But the broom comes up next year in double quantity; for it ſprings from young roots, tho' not from old.

Sheep, fond of broom, devour greedily every young ſhoot; and when they feed on it alone, it is apt to make them drunk, which appears from ſome of them tumbling over when heated by being driven beyond their ordinary pace. In a paſture-field, a few buſhes make a ſhelter; and the cropping of ſheep prevents them in a good meaſure from ſpreading. This ſuggeſts a method of rooting out the young broom that grows in paſture-ground, which is to paſture the field with ſheep. If any eſcape the firſt year, there will not be left a veſtige after the ſecond. Beſide the eaſineſs of this method, it is profitable, as there

there is no food more nouriſhing to ſheep than young broom in moderation.

Where it is neceſſary or convenient to keep the field in corn, horſe hoeing is the only way to ſubdue broom. This ſort of huſbandry encourages every ſeed to vegetate; and the frequent plowings that are neceſſary to a drilled crop, deſtroy broom as well as other weeds. Some may eſcape the firſt year, but they cannot eſcape ſubſequent horſe-hoeing.

A whin is a fine evergreen ſhrub, carrying a ſweet ſmelling flower all the year, except in froſt. But in huſbandry, beauty is not regarded in oppoſition to profit. This ſhrub ſpreads wonderfully in poor ſoil; and when once eſtabliſhed, it is not to be extirpated but by much labour. The roots of broom puſh downward; and for that reaſon, probably, broom ſprings not from the root. The roots of a whin, on the contrary, puſh horizontally, to the diſtance ſometimes of ten or twelve feet. While the plant is growing, it draws the whole juices from the roots; but when cut down, theſe juices puſh up ſuckers from every root that lies near the ſurface. And hence the difficulty of extirpating whins.

The beſt method is, to ſet fire to the whins in a windy day during froſt. Froſt has the effect to wither whins, and to make them burn readily; the harder the froſt, the better. Cut over the ſtumps with a hatchet, and wait till the ground be ſoftned with rain. At that time, a firm plough with a ſtrong draught, will tear all the roots into ſhreds, which will be brought above ground by a heavy brake. The expence of this operation is ſmall, compared with the common method of rooting out whins with a hoe; and it is the leſs to be grudged, as being done in wet weather, when the plough cannot be otherwiſe employed. Oxen are the fitteſt for this operation; becauſe they make a ſteady draught, and do not, like horſes, give up when they meet with unexpected reſiſtance.

If the field be ſoon laid down in graſs, whins will ſpring up in abundance, not only from ſeed, but from the ſmall fibres that may have eſcaped the plough and brake. Paſturing with ſheep is the only remedy: and it is an effectual remedy, if applied immediately after the whins ſpring; for ſheep are no leſs fond of whins than of broom; and if there be a ſufficient number, they will not leave a ſingle plant above ground. Their ſpringing for years need not be repined at; for in their infant ſtate they are excellent food, equal to any plant that grows. If this method be neglected till the whins have grown into wood, ſheep come too late; they will feed on the tender parts, but the buſh continues healthy; of which we have examples every where in open fields.

But if graſs be not immediately wanted, horſe-hoeing huſbandry is the moſt effectual method for

for clearing a field from whins. And when a field is once enriched by labour and manure, one need not be afraid of whins, as a poor ſoil is only proper for them.

Whins are a food ſo nouriſhing, not only to ſheep but to horned cattle, that a poor ſoil cannot be better employed than in bearing crops of them: they are ate pleaſantly in their infant-ſtate, while the prickles are yet ſo tender as not to hurt the mouth.

Having touched the article of food, though foreign to the preſent chapter, I am excited to obſerve, that the tops of the oldeſt whins, cut off with a hedge-bill, make a hearty food for cattle when boiled. And to encreaſe the doze, thiſtles, every bulky weed, and the roots of cabbage, may be boiled with the whins. This I am informed is an ordinary practice in Germany.

Molehills may be juſtly conſidered as an obſtruction to cropping. It is therefore beneficial to deſtroy moles; and the ſimpleſt way is, to lay hold of the young, which are always found in the large molehills. Buffon is uncertain whether they breed more than once a-year; but ſays, that young moles are found from April to Auguſt. What I found, were generally between the firſt of May new ſtyle and the firſt of May old ſtyle.

2. BRING-

2. Bringing into Culture Land from the State of Nature.

The following method of improving a moor is contained in a letter addreſſed to me by an expert farmer. " I began with plowing as much of " the moor as I had a proſpect of dung for. It " was croſs-plowed next ſpring, and well har- " rowed. About Midſummer, it was plowed " again in the direction of the firſt plowing, and " well harrowed. It was plowed the fourth " time before winter, but not harrowed. It got " the fifth plowing in the ſucceeding April, at " which time the ſward was well broken and rot- " ted. It was dunged liberally; and after plow- " ing the dung into the ground, I ſowed turnip. " New ground, however poor originally, never " diſappointed me of a rich crop of turnip the firſt " year, barley or oats the ſecond, and clover and " rye-graſs the third. And yet there are ſeveral " late inſtances of turnip failing in the beſt infield " grounds, among my neighbours and in my own " farm. In the year 1759, unwilling to truſt the " whole of my turnip-crop to the moor, I prepar- " ed and dunged part of my beſt infield, and ſow- " ed it with turnip. Not a ſingle drill appeared " from the firſt ſowing; and but a ſcanty crop " from the third: in the moor not a ſingle drill " failed. I was anxious to diſcover the cauſe:

" I

" I could not attribute the difference to ſoil or
" ſituation, which were in favour of the infield;
" and the management was the ſame. The black
" fly, it is true, had deſtroyed the infield crop;
" but I was puzzled why they had abſtained from
" the moor. Upon examining narrowly the two
" fields, I diſcovered that the black flies in the
" infield were young and very ſmall, not able to
" fly, but only to creep or hop about. Not a
" ſingle fly could be ſeen in the moor. This led
" me to the following ſpeculation. Nature has
" taught the various tribes of flies and inſects, to
" depoſit their eggs in places the beſt adapted for
" the production and nouriſhment of their young.
" Warm and rich grounds anſwer theſe purpoſes
" the beſt: moor-grounds are cold and barren.
" I reflected further, that when once theſe inſects
" chuſe for their habitation a rich ſoil where there
" is plenty of food, they depoſit their eggs there;
" and the race is propagated from year to year.
" A vacuity of ten or twelve yards in one of the
" moor-drills ſeemed not to ſquare with my theo-
" ry. But upon enquiry I found, that a ſmall
" ſhower of rain had ſtopped the holes of the
" drill-box a little time."

So far my correſpondent. His method is good; but it is laborious; and the ſoil with three plowings may be rendered as proper for turnip, as with the ſix he gave it. The following alteration is what I propoſe. Let the moor be opened

in

in winter when it is wet; which has one convenience, that the plough cannot be employed at any other work. In ſpring, after froſt is over, a ſlight harrowing will fill up the ſeams with mould, to keep out the air and rot the ſod. In that ſtate let it lie the following ſummer and winter, which will rot the ſod more than if laid open to the air by plowing. Next April, let it be croſs-plowed, braked, and harrowed, till it be ſufficiently pulverized. Let the manure laid upon it, whether lime or dung, be intimately mixed with the ſoil by repeated harrowings. This will make a fine bed for the turnip-ſeed, if ſown broad-caſt. But if drills be intended, the method muſt be followed that is directed afterward in treating of the culture of turnip.

Repeated experience has convinced me, that turnip in mooriſh ground runs no riſk from the black fly, nor probably in any ground newly broken up. It is an old obſervation of gardeners, that their plants thrive beſt in new ground; for what other reaſon, than that they have no enemies to deſtroy them? It muſt give pleaſure to thoſe who wiſh proſperity to their country, to be informed, that mooriſh grounds of a tolerable ſoil are fit for producing turnip; a profitable crop of itſelf, and ſtill more profitable by preparing the ground for ſubſequent crops. Conſidering the immenſe quantity of moor in Scotland, it muſt give pain to think how little progreſs is made in

its

its cultivation. We ſee a fair example given near twenty years ago, neither expenſive nor intricate. Why does not every farmer exert his utmoſt induſtry upon ſo valuable an improvement? Cuſtom fetters men in chains. To break looſe from ſlavery, a man muſt be bleſſed by nature with a ſuperior degree of underſtanding and activity. Such men there are, though rare: their example will be imitated; and it is a pleaſing proſpect, that our barren moors will in time be converted into good ſoil, productive of nouriſhment for man and beaſt: villages will ariſe, and population go on in a rapid courſe. I preſent to the view of my reader an immenſe moor between Greenlaw in Berwickſhire, and Fala in Mid Lothian, as a deſirable ſubject for an improving farmer, now that there is acceſs to lime by a turnpike-road. As the ſoil for the moſt part is too ſhallow for paring and burning, it may be cultivated according to the directions given above. At the ſame time, there are many ſwampy ſpots, which upon paring and burning will yield a great quantity of aſhes, to be laid upon the dry parts; and theſe aſhes make the very beſt manure for turnip. A ſucceſsful turnip-crop, fed on the ground with ſheep, is a fine preparation for laying down a field with graſs-ſeeds. And it is a ſtill greater improvement, to take two or three ſucceſſive crops of turnip, which will require no dung for the ſecond and following crops. This will thicken the ſoil,

and

and enrich it greatly. In that high country, where rain ſuperabounds, the profitable crop is graſs, not corn. And farms in that country, after being improved, ought to be divided into incloſures of fifty or ſixty acres. Sheep require a hundred acres, to give them ſpace for an extenſive walk; and in ſuch an incloſure any fence keeps them in *.

The beſt way of improving ſwampy ground after draining, is paring and burning. But where the ground is dry, and the ſoil too thin for paring, the beſt way of bringing it into tilth, is to plow it with a feathered ſock, laying the graſſy ſurface under. After the new ſurface is mellowed with froſt, fill up all the ſeams by harrowing croſs the field, which by excluding the air will effectually rot the ſod. In this ſtate let it lie ſummer and winter. In the beginning of May after, a croſs plowing will reduce all to ſmall ſquare pieces, which muſt be pulverized with the brake, to make it ready for a May or June crop. If theſe ſquare

* Scotland probably was moſtly covered with trees before the commencement of agriculture; from which we ought to expect its ſurface to be generally vegetable ſoil. Yet moſt of it is barren earth, without any mixture of vegetable ſoil. Did ſuch large tracts never carry trees? I can hardly think ſo, as trees are found growing in the moſt barren ſpots. Yet in America, which abounds with trees, one in ſome places may travel ſeveral hundred miles without ſeeing a tree. This appears difficult to be accounted for.

ſquare pieces be allowed to lie long in the ſap without brakeing, they will become tough and not be eaſily reduced.

3. FORMING RIDGES.

THE firſt thing that occurs on this head, is to conſider what grounds ought to be formed into ridges, what ought to be tilled with a flat ſurface. Dry ſoils, which ſuffer by lack of moiſture, ought to be tilled flat, in order to retain moiſture. And the method for ſuch tilling, is to go round and round from the circumference to the centre, or from the centre to the circumference. This method is advantageous in point of expedition, as the whole is finiſhed without once turning the plough. At the ſame time, every inch of the ſoil is moved, inſtead of leaving either the crown or the furrow unmoved, as is commonly done in tilling ridges. Clay ſoil, which ſuffers by water, ought to be laid as dry as poſſible by proper ridges. A loamy ſoil is the middle between the two. It ought to be tilled flat in a dry country, eſpecially if it incline to the ſoil firſt mentioned. In a moiſt country, it ought to be formed into ridges, high or low according to the degree of moiſture, and tendency to clay.

In grounds that require ridging, an error prevails, that ridges cannot be raiſed too high. High ridges labour under ſeveral diſadvantages. The ſoil is heaped upon the crown, leaving the furrows bare: the crown is too dry, and the furrows too wet: the crop, which is always beſt on the crown, is more readily ſhaken with the wind, than where the whole crop is of an equal height: the half of the ridge is always covered from the ſun, a diſadvantage which is far from being ſlight in a cold climate. High ridges labour under another diſadvantage in ground that has no more level than barely ſufficient to carry off water: they ſink the furrows below the level of the ground; and conſequently retain water at the end of every ridge. The furrows ought never to be ſunk below the level of the ground. Water will more effectually be carried off, by contracting the ridges both in height and breadth: a narrow ridge, the crown of which is but eighteen inches higher than the furrow, has a greater ſlope, than a very broad ridge where the difference is three or four feet.

Next of forming ridges where the ground hangs conſiderably. Ridges may be too ſteep as well as too horizontal; and if to the ridges be given all the ſteepneſs of the field, a heavy ſhower may do irreparable miſchief. One inſtance I was witneſs to. A hanging field had been carefully dreſſed with lime and dung for turnip. The turnip was fairly above ground, when a fatal ſummer-ſhower ſwept down turnip, lime, dung, and a quantity of the looſe ſoil, leaving the land bare. To prevent ſuch miſchief, the ridges ought

ought to be fo directed crofs the field, as to have a gentle flope for carrying off water flowly, and no more. In that refpect, a hanging field has greatly the advantage of one that is nearly horizontal; becaufe in the latter, there is no opportunity of a choice in forming the ridges. A hill is of all the beft adapted for directing the ridges properly. If the foil be gravely, it may be plowed round and round, beginning at the bottom and afcending gradually to the top in a fpiral line. This method of plowing a hill, requires no more force than plowing on a level: at the fame time it removes the great inconvenience of a gravelly hill, that of rain going off too quickly; for the rain is retained in every furrow. If the foil be fuch as to require ridges, they may be directed to any flope that is proper. Columella, Book 2. chap. 5. advifes, in plowing a hill the furrows to be directed crofs the hill: which he obferves is much eafier than the plowing it up and down.

In order to form a field into ridges that has not been formerly cultivated, the rules mentioned are eafily put in practice. But what if ridges be already formed, that are either crooked or too high? After feeing the advantage of forming a field into ridges, people were naturally led into an error, that the higher the better. But what could tempt them to make their ridges crooked? I anfwer, Not defign, but the lazinefs of the driver fuffering the cattle to turn too ha-

ftily

ſtily, inſtead of making them finiſh the ridge without turning. There is more than one diſadvantage in this ſlovenly practice. Firſt, the water, kept in by the curve at the end of every ridge, ſours the ground. Next, as a plough has the leaſt friction poſſible in a ſtraight line, the friction muſt be increaſed in a curve, the back part of the mould-board preſſing hard on the one hand, and the coulter preſſing hard on the other. In the third place, the plough moving in a ſtraight line, has the greateſt command in laying the earth over. But where the ſtraight line of the plough is applied to the curvature of a ridge in order to heighten it by gathering, the earth moved by the plough is continually falling back, in ſpite of the moſt ſkilful ploughman.

The inconveniencies of ridges high and crooked are ſo many, that one would be tempted to apply a remedy at any riſk. And yet, if the ſoil be clay, I would not adviſe a tenant to apply the remedy upon a leaſe ſhorter than two nineteen years. In a dry gravelly ſoil, the work is not difficult nor hazardous. When the ridges are cleaved two or three years ſucceſſively in the courſe of cropping, the operation ought to be concluded in one ſummer. The earth, by reiterated plowings, ſhould be accumulated upon the furrows, ſo as to raiſe them higher than the crowns: they cannot be raiſed too high, for the accumulated earth will ſubſide by its own weight.

Croſs-

Crofs-plowing once or twice, will reduce the ground to a flat furface, and give opportunity to form ridges at will. The fame method brings down ridges in clay foil: only let care be taken to carry on the work with expedition; becaufe a heavy fhower before the new ridges are formed, will foak the ground in water, and make the farmer fufpend his work for the remainder of that year at leaft. In a ftrong clay, I would not venture to alter the ridges, unlefs it can be done to perfection in one feafon.

In the operation of flattening ridges, there is a circumftance that has fcarce ever been attended to. Ridges are evidently an improvement of art; and fo is manure. The foil muft have been exhaufted by frequent cropping before either ridges or dung were attempted; and after ridges were formed, the foil under the crowns muft continue in its exhaufted ftate, being deprived of the benefit both of fun and rain. We find thefe conjectures verified by experience. When a ridged field is made level as originally, the foil immediately under the crown, which had the thickeft covering, is commonly exceedingly poor. Farmers, fenfible of this, never fail to give more dung to that part than to the reft of the field. But a more effectual remedy is, to pierce deeper than the original furface with the fpade or plough, in order to bring up virgin foil. This foil is generally good; and the whole is made equally fertile, by mixing

mixing it with the exhausted soil in cross-plowing and harrowing. In the low country of Berwickshire, I frequently used this remedy, and never was disappointed. In making ditches there for draining, I have brought up a brick-clay lying four feet under the surface, which without being mellowed either by sun or frost, carried oats of a surprising size. Wild plants may be seen every where much larger than ordinary, growing even in gravel thrown up from the bottom of a new-made ditch*.

Let

* The instructions contained in this work regard a peopled country where every acre is occupied and pays rent. They cannot have place, where the corn-land, by paucity of inhabitants, bears no proportion to what is left waste; which was the case in Scotland two centuries ago, and is at present the case of North America. The prospect of dung will never engage a farmer to submit to the expence of feeding his cattle at home, when they can be maintained in the common without expence. Having thus little or no dung, the farmer crops field after field, till they can bear no more. He chuses fields that are dry; and when these are exhausted, he must attempt moist ground where ridges are necessary. Ridges thus introduced, were, from imitation and custom, extended to all grounds wet and dry. Dry fields exhausted by cropping without manure, were in some measure benefited by ridges, which brought up new soil to the surface, and thickened the crown of the ridge. But soil, rich or poor, when covered from the sun and rain by a thick coat of earth, will, as observed in the text, remain in that state for ever, without turning better or worse. The division of a farm into infield and outfield,

once

Let it be a rule, to direct the ridges north and south if the ground permit. In this direction, the east and west sides of a ridge, dividing the sun equally between them, will ripen at the same time.

It is a great advantage in agriculture, to form ridges so narrow, and so low, as to admit the crowns and furrows to be changed alternately every crop. The soil nearest the surface is the best; and by such plowing, it is always kept near the surface, and never buried. In high ridges, the soil is accumulated at the crown, and the furrows left bare. Such alteration of crown and furrow, is easy where the ridges are no more but seven or eight feet broad. This mode of plowing answers perfectly well in sandy and gravelly soils, and even in loam. But it is not safe in clay soil. In that soil, the ridges ought to be twelve feet wide and twenty inches high; to be preserved always in the same form by casting, that is, by plowing two ridges at a time, beginning at the furrow that separates them, plowing round that furrow, and so round and round till the two ridges be finished. By this method, the separating furrow is raised a little higher than the furrows that bound the two ridges. But at the next

once universal in Scotland, and still frequent, proceeds from the same cause. We have therefore no reason to blame our forefathers for a practice perfectly well suited to a a country slack of inhabitants.

next plowing, that inequality is corrected, by beginning at the bounding furrows, and going round and round till the plowing of the two ridges be completed at the separating furrow.

I cannot conclude this article without inveighing against the commons in England, which produce very little and are destructive by increasing the number of the idlers who subsist by charity. An act for dividing them, as in Scotland, would give work to many thousands, would increase population, would add greatly to the land product, and no less to the public revenue.

4. Clearing Ground of Weeds.

The farmer views plants in a very different light from the botanist. All are weeds with the farmer that give obstruction to the plants he propagates in his farm. These I distinguish into two kinds, that require different management, *viz.*—first, annuals; and next, all that have a longer existence, which I shall comprehend under the general name of *perennials*. It is vain to expect a crop of corn from land over-run with couch-grass, knot-grass, or other perennial weeds; and yet the time may be remembered, when, among Scotch farmers, it was a disputed point, whether such weeds were not more profitable than hurtful. Some found them profitable in binding their light land: the getting a plentiful crop of straw and

and hay for their cattle, weighed with others. I ſhould be aſhamed of expoſing ignorance ſo groſs in my countrymen, could I not ſay, that they now underſtand the matter better, though few of them hitherto have arrived at the perfection of cleaning. Summer-fallow is the general method; and excellent it is, though it doth not always prove effectual. The roots of couch-graſs in particular are long, and full of juice: if a ſingle joint be left in the ground, it never fails to ſpring. Here the common harrow is of very little uſe, its teeth being too wide. The time relied on by our farmers for deſtroying couch-graſs, is in prepa-ing for barley. After the harrow has raiſed part of a root above ground, men, women, and children, are employed to pull it up. There are only two objections to this method: the expence is one; the other is, that after all this expence, many roots are left in the ground. In order to pave the way for rooting out perennials effectually, and with little expence, I take liberty to introduce a new inſtrument, which I term a *cleaning harrow*. It is of one entire piece, like the firſt of thoſe mentioned above, conſiſting of ſeven bulls, four feet long each, two and one fourth inches broad, two and three fourths deep. The bulls are united together by croſs bars, ſimilar to what are mentioned above. The intervals between the bulls being three and three fourths inches, the breadth of the whole harrow is three feet five inches. In each bull are inserted eight teeth, each nine inches free below the wood, and distant from each other six inches. The weight of each tooth is a pound, or near it. The whole is firmly bound by an iron plate from corner to corner in the line of the draught. The rest as in the harrows mentioned above. The size, however, is not invariable. The cleaning harrow ought to be larger or less according as the soil is stiff or free. See the figure annexed.

To give this instrument its full effect, stones of such a size as not to pass freely between the teeth, ought to be carried off, and clods of that size ought to be broken. The ground ought to be dry, which it is commonly in the month of May.

In preparing for barley, turnip, or other summer-crop, begin with plowing and cross-plowing. If the ground be not sufficiently pulverized, let the great brake be applied, to be followed successively with the harrows, No. 1. and No. 2. In stiff soil rolling may be proper, once or twice between the acts. These operations will loosen every root, and bring some of them to the surface. This is the time for the harrow, No. 3. conducted by a boy mounted on one of the horses, who trots smartly along the field, and brings all the roots to the surface: there they are to lie for a day or two till perfectly dry. If any stones or clods remain, they must be carried off

in

in a cart. And now fucceeds the operation of the cleaning harrow. It is drawn by a fingle horfe, directed by reins, which the man at the oppofite corner puts over his head, in order to have both hands free. In this corner is fixed a rope, with which the man from time to time raifes the harrow from the ground, to let the weeds drop. For the fake of expedition, the weeds ought to be dropt in a ftraight line crofs the field, whether the harrow be full or no; and feldom is a field fo dirty but that the harrow may go thirty yards before the teeth are filled. The weeds will be thus laid in parallel rows, like thofe of hay raked together for drying. A harrow may be drawn fwiftly along the rows, in order to fhake out all the duft; and then the weeds may be carried clean off the fields in carts. But we are not yet done with thefe weeds: inftead of burning, which is the ordinary practice, they may be converted into ufeful manure, by laying them in a heap with a mixture of hot dung to begin fermentation. What better politic than to make a friend of a foe! At firft view, this way of cleaning land will appear operofe; but upon trial, neither the labour nor the expence will be found immoderate. So far from it, that I believe it will not be eafy to name any other way of cleaning ground effectually that cofts fo little. At any rate, the labour and expence ought not to be grudged; for if a field be once thoroughly

thoroughly cleaned, the ſeaſons muſt be very croſs, or the farmer very indolent, to make it neceſſary to renew the operation in leſs than twenty years *.

Moſs is one of the moſt pernicious weeds that enter into a graſs-field: in a very dry ſoil, it uſurps upon the good plants, wears them out, and covers the whole ſurface. I have tried lime and dung to very little purpoſe. Coal-aſhes do better: they keep it back a few years; but it recovers ſtrength gradually as the aſhes loſe their influence, and prevails as much as before. Some writers talk of rolling, becauſe moſs is never ſeen on a foot-path. They do not advert, that the continual treading of feet on a narrow path, conſolidates the ground more, and makes it more retentive

* Since the laſt edition, a letter recieved from a gentleman no leſs accurate than ſkilful in huſbandry, is in the following words: "The ground preparing for turnip was "very ſoul. I cauſed harrow it with common harrows "as well as poſſible, and employed people with rakes to "gather the wreck: When the common harrows could "do no more ſervice, I employed your cleaning harrow, "which from ſour acres and a half brought to the ſurface "as much wreck as loaded ten ſingle horſe carts. I tried "it again this ſummer on a ſmall bit of ground, about a "quarter of an acre, near my houſe, in order to lay it "down with graſs ſeeds. After the common harrows and "gardeners with rakes, had cleaned it as well as they "could, a quantity was gathered by your cleaning harrow "that filled four of the carts mentioned."

retentive of moiſture, than rolling can do. Nor do they advert, that moſs is the moſt plentiful in dry ground, where the weightieſt roller makes no impreſſion. As moſs proſpers the moſt on the drieſt ground, the laying the field under water a whole winter, will deſtroy it. There is reaſon however to ſuſpect, that it will encroach again, when the moiſture is exhauſted, and the ſoil returns to its arid ſtate; not to mention the many fields that lie out of the reach of water. One infallible method there is; which is, to cover the ground an inch thick with ſoil retentive of moiſture, and to mix it with the original ſoil by the plough and harrow. Thus will moſs be baniſhed, and the ſoil at the ſame time enriched. But the good ſoil here muſt be at hand: it would be too expenſive to bring it from a diſtance. Lucky it is, after all, that there ſtill remains an infallible method, which, inſtead of being expenſive, is extremely profitable. It will be made evident afterward, in laying down rules for rotation of crops, that a quick ſucceſſion of corn and graſs, is more profitable, than to allow any ground to continue long in either. Therefore, as ſoon as moſs begins to prevail, plow up the ground for a crop of corn *.

Next,

* The following method is propoſed by Dr Home for deſtroying moſs. "Shut up the incloſure from the middle "of May till the beginning of December: feed it till A-"pril,

Next, of annual weeds, which are propagated by ſeed. As ſeeds cannot be gathered out of the ground like roots, the only way of deſtroying them, is to promote their vegetation; which lays them open to be extirpated by the plough or harrow. For want of induſtry, annual weeds prevail in many parts of this iſland, in the beſt ſoils eſpecially. To view a crop near a town, end of May, or beginning of June, one would believe it to be a crop of charlock or wild muſtard. Theſe plants, it is true, loſe their ſplendid appearance after their flowering is over; but they remain to encumber the ground. As they ripen, and drop their ſeed long before the corn is ripe, they multiply more and more. They not only rob the ground, and ſtarve the good grain, but prevent circulation of air about the roots, which is a great impediment to vegetation. Freedom of circulation is one of the cauſes that make drilled crops ſucceed ſo well. What muſt have been the condition of corn-land in Scotland before fallowing was known!

And

" pril, and then ſave it for a crop of hay. The moſs be-
" ing ſo long covered with graſs, is cut off from the bene-
" fit of the air, and dies." I doubt. Over-ſhadowing may retard its progreſs; but as over-ſhadowing does not correct the dryneſs of the ſoil, it is not likely to complete the cure. A haugh on the river Nith was over-run with moſs, and lime proved a cure. But moſs growing on a rich haugh at the ſide of a river muſt be different from what grows on a dry gravelly ſoil that holds no water.

And now to the deſtruction of ſuch weeds. Summer-fallow, among its other advantages, does this effectually, by obſerving the following method. Begin with plowing in April as ſoon as the ground is dry; and let the brake ſucceed croſs the field; which by pulveriſing the ſoil, will promote the vegetation of every ſeed. As ſoon as the weeds appear, which may be in ten or twelve days according to the ſeaſon, plow and brake as before: the plough will make many ſeeds vegetate; the brake ſtill more. Proceed in the ſame courſe while any weeds appear. In the heat of ſummer, rolling not only promotes vegetation by keeping in the moiſture, but bruiſes the clods which lock up many ſeeds.

This proceſs requires the following precaution. Avoid plowing or brakeing when the ground is wet. The ſtirring wet ground hardens it, excludes the ſun and air, and prevents vegetation.

CHAP. V.

CULTURE OF PLANTS FOR FOOD.

THE articles hitherto inſiſted on, are all of them preparatory to the capital object of a farm, that of raiſing plants for the nouriſhment of man, and of other animals. Theſe are of two

two kinds; culmiferous, and leguminous, differing widely from each other. Wheat, rye, barley, oats, ryegrafs, are of the firft kind: of the other kind are peafe, beans, clover, cabbage, and many others. Culmiferous plants, fays Bonnet, have three fets of roots. The firft iffue from the feed, and pufh to the furface an upright ftem; another fet iffue from a knot in that ftem; and a third, from another knot, nearer the furface. Hence the advantage of laying feed fo deep in the ground as to afford fpace for all the fets. Leguminous plants form their roots differently. Peafe, beans, cabbage, have ftore of fmall roots, all iffuing from the feed, like the undermoft fet of culmiferous roots; and they have no other roots. A potato and a turnip have bulbous roots. Red clover has a ftrong tap-root. The difference between culmiferous and leguminous plants with refpect to the effects they produce in the foil, will be explained afterward, in the chapter concerning rotation of crops. As the prefent chapter is confined to the propagation of plants, it falls naturally to be divided into three fections: Firft, plants cultivated for fruit; Second, plants cultivated for roots; Third, plants cultivated for leaves.

SECTION I.

PLANTS CULTIVATED FOR FRUIT.

WHEAT, rye, oats, barley, beans, and peaſe, are the plants that are moſtly cultivated in Scotland for fruit. I begin with wheat and rye.

1. WHEAT and RYE.

As ſoon in ſpring as the ground is fit for plowing, the fallowing for wheat may commence. The moment ſhould be choſen, when the ground, beginning to dry, has yet ſome remaining ſoftneſs: in that condition, the ſoil divides eaſily by the plough, and falls into ſmall parts. This is an eſſential article, deſerving the ſtricteſt attention of the farmer. Ground plowed too wet, riſes, as we ſay, *whole-fur*, as when paſture-ground is plowed: where plowed too dry, it riſes in great lumps, which are not reduced by ſubſequent plowings; not to mention, that it requires double force to plow ground too dry, and that the plough is often ſhattered. When the ground is in proper order, the farmer can have no excuſe for delaying a ſingle minute. This firſt courſe of fallow, muſt, it is true, yield to the barley-ſeed; but as the barley-ſeed is commonly over the firſt week of May, or ſooner, the ſeaſon muſt be unfavourable if the fallow cannot be reached by the middle of May.

As clay ſoil requires high ridges, theſe ought to be cleaved at the firſt plowing, beginning at the furrow, and ending at the crown. This plowing ought to be as deep as the ſoil will admit; and water-furrowing ought inſtantly to follow: for if rain happen before water-furrowing, it ſtagnates in the furrow, neceſſarily delays the ſecond plowing till that part of the ridge be dry, and prevents the furrow from being mellowed, and roaſted by the ſun. If this firſt plowing be well executed, annual weeds will riſe in plenty. About the firſt week of June, the great brake will looſen and reduce the ſoil, encourage a ſecond crop of annuals, and raiſe to the ſurface the roots of weeds moved by the plough. Give the weeds time to ſpring, which may be in two or three weeks. Then proceed to the ſecond plowing about the beginning of July; which muſt be croſs the ridges, in order to reach all the ſlips of the former plowing. By croſs-plowing, the furrows will be filled up, and water-furrowing be ſtill more neceſſary than before. Employ the brake again about the 10th of Auguſt, to deſtroy the annuals that have ſprung ſince the laſt ſtirring. The deſtruction of weeds is a capital article in fallowing: yet ſo blind are people to their intereſt, that nothing is more common, than

a

a fallow field covered with charlock and wild muſtard, all in flower, and ten or twelve inches high. The field having now received two harrowings and two brakeings, is prepared for manure, whether lime or dung, which without delay ought to be incorporated with the ſoil, by repeated harrowing and a gathering furrow. This ought to be about the beginning of September; and as ſoon after as you pleaſe the ſeed may be ſown.

As in plowing a clay ſoil it is of importance to prevent poaching, the hinting furrows ought to be done with two horſes in a line. If four ploughs be employed in the ſame field, to one of them may be allotted the care of the hinting furrows. What will one think, who has never been in the Highlands, of four horſes abreaſt in a plough, the driver going backward, and ſtriking the horſes on the forehead to make them come forward? The ignorant and illiterate are ſtrangely dull in point of invention: let the moſt ſtupid practice turn once cuſtomary, and it is rivetted in them for ever. Is it not obvious, that four horſes abreaſt muſt tread down the new-moved ſoil, and reduce it to as firm a ſtate as before the plough entered?

Next of dreſſing loam for wheat. Loam, being a medium between ſand and clay, is of all ſoils the fitteſt for culture, and the leaſt ſubject to chances. It does not hold water like clay;

and

and when wet, is ſooner dry. At the ſame time, it is more retentive than ſand of that degree of moiſture which promotes vegetation. On the other hand, it is more ſubject to couch-grals than clay, and to other weeds; to deſtroy which, fallowing is ſtill more neceſſary than in clay.

Beginning the fallow about the firſt of May, or as ſoon as barley-ſeed is over, take as deep a furrow as the ſoil will admit. Where the ridges are ſo low and narrow as that the crown and furrow can be changed alternately, there is little or no occaſion for water-furrowing. Where the ridges are ſo high as to make it proper to cleave them, water-furrowing is proper. The ſecond plowing may be at the diſtance of five weeks. Two crops of annuals may be got in the interim, the firſt by the brake, and the next by the harrow; and by the ſame means eight crops may be got in the ſeaſon. The ground muſt be cleared of couch-graſs and knot-graſs roots, by the cleaning harrow, deſcribed above. The time for this operation is immediately before the manure is laid on. The ground at that time being in its looſeſt ſtate, parts with its graſs-roots more freely than at any other time. After the manure is ſpread, and incorporated with the ſoil by brakeing or harrowing, the ſeed' may be ſown under furrow if the ground hang ſo as eaſily to carry off the moiſture. To leave it rough without harrowing, has two advantages; it is not apt to cake

cake with moiſture; and the inequalities make a ſort of ſhelter to the young plants againſt froſt. But if it lie flat, it ought to be ſmoothed with a ſlight harrow after the ſeed is ſown, which will facilitate the courſe of the rain from the crown to the furrow.

A ſandy ſoil is too looſe for wheat. The only chance for a crop is after red clover, the roots of which bind the ſoil; and the inſtructions above given for loam are applicable here. Rye is a crop much fitter for ſandy ſoil than wheat; and like wheat it is generally ſown after a ſummer-fallow.

Laſtly, Sow wheat as ſoon in the month of October as the ground is ready. When ſown a month more early, it is too forward in the ſpring, and apt to be hurt by froſt: when ſown a month later, it has not time to root before froſt comes on, and froſt ſpews it out of the ground.

2. OATS.

As winter-plowing enters into the culture of oats, I muſt remind the reader of the effect of froſt upon tilled land. Providence has neglected no region intended for the habitation of man. If in warm climates the ſoil be meliorated by the ſun, it is no leſs meliorated by froſt in cold climates. Froſt acts upon water, by expanding it and

and making it to occupy a larger ſpace. Froſt has no effect upon dry earth; witneſs ſand upon which it makes no impreſſion. But upon wet earth it acts moſt vigorouſly: it expands the moiſture, which puts every particle of the earth out of its place, and ſeparates them from each other. In that view, froſt may be conſidered as a plough ſuperior to any that can be made by the hand of man: its action reaches the minuteſt particles; and, by dividing and ſeparating them, it renders the ſoil looſe and friable. This operation is the moſt remarkable in tilled land, which gives free acceſs to froſt. With reſpect to clay ſoil in particular, there is no rule in huſbandry more eſſential than to open it before winter in hopes of froſt. It is even adviſable in a clay ſoil to leave the ſtubble rank, which, when plowed in before winter, keeps the clay looſe, and admits the froſt into every cranny.

To apply this doctrine, it is dangerous to plow clay ſoil when wet; becauſe water is a cement for clay, and binds it ſo as to render it unfit for vegetation. It is however leſs dangerous to plow wet clay before winter than after. A ſucceeding froſt corrects the bad effects of ſuch plowing: a ſucceeding drought encreaſes them. No rule is ſo eaſy to be followed as what I am inculcating; and yet no other is ſo frequently tranſgreſſed. Many farmers have a ſort of baſtard induſtry,

induſtry, that prompts activity without ever thinking of conſequences.

And now to the culture of oats, a culmiferous plant. The common method is, to ſow them on new-plowed land in the month of March, as ſoon as the ground is tolerably dry. If it continue wet all the month of March, it is too late to venture them after. It is much better to ſummer-fallow and to ſow wheat in the autumn. But the preferable method, eſpecially in clay ſoil, is to turn over the field after harveſt, and to lay it open to the influences of froſt and air, which leſſen the tenacity of clay, and reduce it to a free mould. The ſurface-ſoil by this means is finely mellowed for reception of the ſeed; and it would be a pity to bury it by a ſecond plowing before ſowing. In general, the bulk of clay ſoils are rich; and ſkilful plowing without dung, will probably give a better crop than unſkilful plowing with dung.

Hitherto of natural clays. I muſt add a word of carſe clays which are artificial, whether left by the ſea, or ſweeped down from higher grounds by rain*. The method commonly uſed of dreſſing carſe clay for oats, is, not to ſtir it till the ground be dry in the ſpring, which ſeldom happens before the firſt of March; and the ſeed is ſown as ſoon after as the ground is ſufficiently dry for its reception. Froſt has a ſtronger effect on ſuch clays than on natural clay. And if the field be

* See carſe clay deſcribed Part II. chap. 1. § 1.

be laid open before winter, it is rendered ſo looſe by froſt as freely to admit rain. The particles at the ſame time are ſo ſmall, as that the firſt drought in ſpring makes the ſurface cake or cruſt. The difficulty of reducing this cruſt into mould for covering the oat-ſeed, has led farmers to delay plowing till the month of March. But we are taught by experience, that this ſoil plowed before winter, is ſooner dry than when the plowing is delayed till ſpring; and as early ſowing is a great advantage, the objection of the ſuperficial cruſting is eaſily removed by the harrow, No. 1. above deſcribed, which will produce abundance of mould for covering the ſeed. The plowing before winter not only procures early ſowing, but has another advantage: the ſurface-ſoil that had been mellowed during winter by the ſun, froſt, and wind, is kept above. I have no experience of managing carſe clay in this manner; but to me it appears greatly preferable to the common practice. One accurate experiment I am informed of, that juſtifies my opinion. A carſe field was cleaved in October as a preparation for the next year's fallow. One ridge happened to be left, which was not tilled till the beginning of March following. At the end of that month after a fall of rain, the early tilled ground was dry, and the ridge very wet.

The dreſſing a loamy ſoil for oats, differs little from dreſſing a clay ſoil, except in the following particular,

particular, that being leſs hurt by rain, it requires not high ridges; and therefore ought to be plowed crown and furrow alternately.

Where there is both clay and loam in a farm, it is obvious from what is ſaid above, that the plowing of the clay after harveſt, ought firſt to be diſpatched. If both cannot be overtaken that ſeaſon, the loam may be delayed till the ſpring with leſs hurt.

Next of a gravelly ſoil; which is the reverſe of clay, as it never ſuffers but from want of moiſture. Such a ſoil ought to have no ridges; but plowed circularly from the centre to the circumference, or from the circumference to the centre. It ought to be tilled after harveſt; and the firſt dry weather in ſpring ought to be laid hold of to ſow, harrow, and roll; which will preſerve it in ſap. One uſed to ridges may find ſome difficulty in ſowing without them. But a proper breadth may be marked, either with poles or by the ſower's confining himſelf to a certain number of the circular furrows.

The culture of oats is the ſimpleſt of all. That grain is probably a native of Britain: it will grow on the worſt ſoil with very little preparation. For that reaſon, before turnip was introduced, it was always the firſt crop upon land broken up from the ſtate of nature.

Upon ſuch land, may it not be a good method, to build up on the crown of every ridge, in the

N form

form of a wall, all the ſurface earth, one ſod above another, as in a fold for ſheep? After ſtanding in this form all the ſummer and winter, let the walls be thrown down, and the ground prepared for oats. This will ſecure one or two good crops; after which the land may be dunged for a crop of barley and graſs-ſeeds. This method may anſwer in a farm where manure is ſcanty.

3. Barley.

Barley is a culmiferous plant that requires a mellow ſoil. Upon that account, extraordinary care is requiſite where it is to be ſown in clay. The land ought to be ſtirred immediately after the foregoing crop is removed, which lays it open to be mellowed with froſt and air. In that view, a peculiar ſort of plowing has been introduced, termed *ribbing*; by which the greateſt quantity of ſurface poſſible is expoſed to air and froſt. The obvious objection to this method is, that half of the ridge is left unmoved. And to obviate that objection, I offer the following method, which moves the whole ſoil, and at the ſame time expoſes the ſame quantity of ſurface to froſt and air. This I eſteem a valuable improvement; and I am only in pain about making it to be clearly underſtood. As ſoon as the former crop is off the field, let the ridges be gathered with as deep a furrow as the ſoil will admit,

mit, beginning at the rown and ending at the furrows. This plowing loofens the whole foil, giving free accefs to air and froft. Soon after, begin a fecond plowing in the following manner. Let the field be divided by parallel lines crofs the ridges, with intervals of thirty feet or fo. Plow once round an interval, beginning at the edges, and turning the earth toward the middle of the interval; which covers a foot or fo of the ground formerly plowed. Within that foot plow another round fimilar to the former; and after that other rounds, till the whole interval be finifhed, ending at the middle. Inftead of beginning at the edges, and plowing toward the middle, it will have the fame effect to begin at the middle and to plow toward the edges. Plow the other intervals, in the fame manner. As this operation will fill up the furrows, let them be cleared and water-furrowed without delay. By this method, the field will be left waving like a plot in a kitchen-garden, ridged up for winter. In this form, the field is kept perfectly dry; for befide the capital furrows that feparate the ridges, every ridge has a number of crofs furrows that carry the rain inftantly to the capital furrows. In hanging grounds retentive of moifture, the parallel lines above-mentioned ought not to be perpendicular to the furrows of the ridges, but to be directed a little downward, in order to carry rain-water the more haftily haſtily to theſe furrows. If the ground be clean, it may lie in that ſtate winter and ſpring, till the time of ſeed-furrowing. If weeds happen to riſe, they muſt be deſtroyed by plowing, or brakeing, or both; for there cannot be worſe huſbandry, than to put ſeed into dirty ground.

This method reſembles common ribbing in appearance, but is very different in reality. As the common ribbing is not preceded by a gathering furrow, the half of the field is left untilled, firm as when the former crop was removed, impervious in a great meaſure to air or froſt. It at the ſame time lodges the rain-water on every ridge, preventing it from deſcending to the furrows; which is hurtful in all ſoils, and poiſonous in a clay ſoil. The ſtitching here deſcribed, or ribbing if you pleaſe to call it ſo, prevents theſe noxious effects. By the two plowings the whole ſoil is opened, admitting freely air and froſt; and the multitude of furrows lays the ſurface perfectly dry, giving an early opportunity for the barley-ſeed.——But I have more to ſay in favour of the method propoſed. When it is proper to ſow the ſeed, all is laid flat with the brake, which is an eaſy operation upon ſoil that is dry and pulverized; and the ſeed-furrow which ſucceeds, is ſo ſhallow as to bury little or none of the ſurface-earth: whereas the ſtirring for barley is commonly done with the deepeſt furrow; and conſequently buries all the ſurface-ſoil that was mellowed

lowed by the froſt and air. Nor is this method more expenſive; becauſe the common ribbing muſt always be followed with a ſtirring furrow, which is ſaved in the method recommended. Nay, it is leſs expenſive; for after common ribbing, which keeps in the rain-water, the ground is commonly ſo ſoured, as to make the ſtirring a laborious work.

Where the land is in good order, and free of weeds, April is the month for ſowing barley. Every day is proper from the firſt to the laſt. But in a light ſoil, the later part of the month is the ſafeſt. Three lippies ſown firſt of April, were cut twelfth of Auguſt. The increaſe was 36 lippies, weighing at the rate of 18 ſtone *per* boll. The ſtraw weighed 17 ſtone and 8 pounds. In the ſame field, all of one ſoil, other three lippies were ſown firſt of May and cut ſixth of September. The increaſe 67 lippies, or one boll and three lippies, weighing as the former did at the rate of 18 ſtone *per* boll. The ſtraw weighed 30 ſtone.

The dreſſing loamy ſoil and light ſoil for barley, is the ſame with that deſcribed; only that to plow dry is not altogether ſo eſſential as in dreſſing clay ſoil. Loam or ſand may be ſtirred a little moiſt: better, however, delay a week or two, than to ſtir a loam when moiſt. Clay muſt never be plowed moiſt, even though the ſeaſon ſhould eſcape altogether. But this will ſeldom be be neceſſary; for not in one year of twenty will it happen, but that clay is dry enough for plowing ſome time in May. Froſt may correct clay plowed wet after harveſt; but plowed wet in ſpring, it unites into a hard maſs, not to be diſſolved but by very hard labour.

The foregoing culmiferous plants are what are ordinarily propagated for food in Scotland. What follow are leguminous plants. And I begin with beans, being ſown the earlieſt in ſpring.

4. Beans.

The propereſt ſoil for beans is a deep and moiſt clay.

There was lately introduced into Scotland a method of ſowing beans with a drill-plough, and horſe-hoeing the intervals; which, beſide affording a good crop, is a dreſſing to the ground. But as that method is far from being general, I keep in the common track.

As this grain is early ſown, the ground intended for it ſhould be plowed before winter, to give acceſs to froſt and air; beneficial in all ſoils, and neceſſary in a clay ſoil. Take the firſt opportunity after January when the ground is dry, to looſen the ſoil with the harrow firſt deſcribed, till a mould be brought upon it. Sow the ſeed, and cover it with the ſecond harrow. The third will ſmooth

ſmooth the ſurface, and cover the ſeed equally. Theſe harrows make the beſt figure in covering beans, which by them can be laid deeper in the ground than by the ordinary harrows: In clay ſoil, the common harrows are altogether inſufficient. The ſoil that has reſted long after plowing, is rendered compact and ſolid: the common harrows ſkim the ſurface: the ſeed is not covered; and the firſt hearty ſhower of rain lays it above ground. Where the farmer overtakes not the plowing after harveſt, and is reduced to plow immediately before ſowing, the plough anſwers the purpoſe of the firſt harrow; and the other two will complete the work. But the labour of the firſt harrow is ill ſaved; as the plowing before winter is a fine preparation, not only for beans, but for grain of every kind. If the ground plowed before winter happen by ſuperfluity of moiſture to cake, the firſt harrow going along the ridges, and croſſing them, will looſen the ſurface, and give acceſs to air for drying. As ſoon as the ground is dry, ſow without delaying a moment. If rain happen in the interim, there is no remedy but patience till a dry day or two come.

Carſe clay, plowed before winter, ſeldom fails to cake. Upon that account, a ſecond plowing is neceſſary before ſowing; which ought to be performed with an ebb furrow, in order to keep the froſt-mould as near the ſurface as poſſible. To cover the ſeed with the plough is expreſſed by the

the phrafe *to fow under furrow.* The clods raifed in this plowing, are a fort of fhelter to the young plants in the chilly fpring-months.

The foregoing method will anfwer for loam. And as for a fandy or gravelly foil, it is altogether improper for beans.

Though I cannot recommend the horfe-hoeing of beans, with the intervals that are commonly allotted for turnip, yet I warmly recommend the drilling them at the diftance of ten or twelve inches, and keeping the intervals clean of weeds. This may be done by hand-hoeing, taking opportunity at the fame time to lay frefh foil to the roots of the plants. But as this is an expenfive operation, and hands are not always to be got, I propofe a narrow plough, drawn by a fingle horfe, with a mouldboard on each fide to lay the earth upon the roots of the plants. This is a cheap and expeditious method: it keeps the ground clean; and nourifhes the plants with frefh foil *.

As

* Cornelius Celfus declares againft weeding or hoeing beans; becaufe, fays he, "after having pulled them up "with the hand, a crop of grafs remains for making hay." I have often regreted the lofs of that author's work upon hufbandry; becaufe, from that on medicine, he appears a firft rate genius. But, if we can truft to a fpecimen, the lofs is not great. Columella, book 2d, ch. 12th, juftly condemns him for this doctrine. "It appears, (fays he,) bad "hufbandry to fuffer weeds to grow up with corn, which "muft deprive the corn of fo much nourifhment." He adds, with refpect to beans in particular, "that the keep-"ing them clean produces much meal and a very thin hufk."

As beans delight in a moiſt ſoil, and have no end of growing in a moiſt ſeaſon, they cover the ground totally when ſown broadcaſt, keep in the dew, and exclude the ſun and air: the plants grow to a great height, but carry little ſeed, and that little not well ripened. This diſplays the advantage of drilling, which gives free acceſs to the ſun and air, dries the ground, and affords plenty of ripe ſeed.

5. PEASE.

PEASE are of two kinds; the white, and the gray. The latter is what generally is cultivated in Scotland for fruit; and the latter only ſhall be here handled, leaving the former to gardeners.

There are two ſpecies of the gray kind, diſtinguiſhed by their time of ripening. One ripens ſoon, and for that reaſon is termed *hot ſeed:* the other, which is ſlower in ripening, is termed *cold ſeed.* Whether theſe be really different ſpecies, or be accidently different only, is left to natural philoſophers.

Peaſe, a leguminous crop, is proper to intervene between two culmiferous crops; leſs for the profit of a peaſe-crop, than for meliorating the ſoil. Peaſe however in a dry ſeaſon will produce ſix or ſeven bolls each acre; but in an ordinary ſeaſon they ſeldom reach above two or two and a half. This leads me to think, that in a moiſt climate, which all the weſt of Britain is, red clover is a more beneficial crop; as it makes as good winter-food as peaſe, and can be cut green thrice during ſummer. When huſbandry comes to be better underſtood, I have little doubt but that red clover will baniſh peaſe altogether, except in the warmeſt and drieſt ſpots.

A field intended for cold ſeed ought to be plowed in October or November; but the ſeed ought not to be put into the ground before March: more early ſowing ſubjects the tender plants to the nipping cold of April. A field intended for hot ſeed, ought to be plowed the middle or end of April immediately before ſowing. But if infeſted with weeds, it ought to be alſo plowed in October or November.

Peaſe laid a foot below the ſurface will vegetate: but the moſt approved depth is ſix inches in light ſoil, and four inches in clay ſoil; for which reaſon, they ought to be ſown under furrow when the plowing is delayed till ſpring. Of all grain, beans excepted, they are the leaſt in danger of being buried.

Peaſe differ from beans, in loving a dry ſoil and a dry ſeaſon. Horſe-hoeing would be a great benefit, could it be performed to any advantage; but peaſe grow expeditiouſly, and ſoon fall over and cover the ground, which bars plowing. Horſe-hoeing has little effect when the plants are new ſprung; and when they are advanced

vanced to be benefited by that culture, their length prevents it. Faſt growing at the ſame time is the cauſe of their carrying ſo little ſeed: the ſeed is buried among the leaves; and the ſun cannot penetrate to make it grow and ripen. For the ſame reaſon, in a peaſe crop, there is always more grain on the weſt ſide of the ridge than on the eaſt ſide. The plants are commonly laid over by the weſt wind, and ſmother the ſeed on the eaſt ſide. The only practicable method to obtain grain, is thin ſowing, but thick ſowing produces more ſtraw, and mellows the ground more. Half a boll for an Engliſh acre may be reckoned thin ſowing; three firlots, thick ſowing.

Notwithſtanding what is ſaid above, Mr Hunter, a noted farmer in Berwickſhire, has begun of late to ſow all his peaſe in drills; and he never fails to have great crops of corn as well as of ſtraw. He ſows double rows with a foot interval, and two feet and a half between the double rows, which admit horſe-hoeing. By that method, he has alſo good crops of beans on light land.

Peaſe and beans mixed are often ſown together, in order to catch different ſeaſons. In a moiſt ſeaſon, the beans make a good crop; in a dry ſeaſon, the peaſe.

The growth of plants is commonly checked by drought in the month of July; but promoted by rain

rain in Auguſt. In July, graſs is parched; in Auguſt, it recovers verdure. Where peaſe are ſo far advanced in the dry ſeaſon as that the ſeed begins to form, the growth of the plants is indeed checked, but the ſeed continues to fill. If the plants are only in the bloſſom at that ſeaſon, their growth is checked a little; but they become vigorous again in Auguſt, and continue growing without filling till ſtopped by froſt. Hence it is, that cold ſeed, which is early ſown, has the beſt chance to produce corn: hot ſeed, which is late ſown, has the beſt chance to produce ſtraw.

The following method is practiſed in Norfolk, for ſowing peaſe upon a dry light ſoil, immediately opened from paſture. The ground is pared with a plough extremely thin, and every ſod is laid exactly on its back. In every ſod a double row of holes is made. A pea dropt in every hole lodges in the flay'd ground immediately below the ſod, thruſts its roots horizontally, and has ſufficient moiſture. This method enabled Norfolk farmers, in the barren year 1740, to furniſh white peaſe to us at 12 s. per boll.

SECT. II.

Plants cultivated for Roots.

Plants of that kind commonly cultivated in the field, are turnip, potatoes, carrot, parſnip.

1. Tur-

1. TURNIP.

IT animates me to have opportunity for giving directions about a crop, that the beſt farmers in this country have now taken into their plan of huſbandry; and that does not altogether eſcape even ſmall farmers. Nor am I acquainted with a ſingle inſtance in Scotland, where turnip fairly begun have been relinquiſhed.

The proper ſoil for turnip is a gravelly ſoil; and there it can be raiſed to the greateſt perfection, and with the leaſt hazard of miſcarrying. At the ſame time, there is no ſoil but will bear turnip when well prepared.

No perſon ever deſerved better of a country, than he who firſt cultivated turnip in the field. No plant is better fitted for the climate of Britain, no plant proſpers better in the coldeſt parts of it, and no plant contributes more to fertility. In a word, there has not for two centuries been introduced into Britain a more valuable improvement.

Of all roots, turnip requires the fineſt mould; and to that end, of all harrows froſt is the beſt. In order to give acceſs to froſt, the land ought to be prepared by ribbing after harveſt, as above directed in preparing land for barley. If the field be not ſubject to annuals, it may lie in that ſtate till the end of May; otherwiſe the weeds

muſt

muſt be deſtroyed by a brakeing about the middle of April; and again in May, if weeds riſe. The firſt week of June, plow the field with a ſhallow furrow. Lime it if requiſite, and harrow the lime into the ſoil. Draw ſingle furrows with intervals of three feet, and lay dung in the furrows. Cover the dung ſufficiently, by going round it with the plough, and forming the three feet ſpaces into ridges. The dung comes thus to lie under the crown of every ridge.

The ſeaſon of ſowing muſt be regulated by the time intended for feeding. Where intended for feeding in November, December, January, and February, the ſeed ought to be ſown from the 1ſt to the 20th of June. Where the feeding is intended to be carried on till March, April, and May, the ſeed muſt not be ſown till the end of July. Turnip ſown earlier than above directed, flowers that very ſummer, and runs faſt to ſeed; which renders it in a good meaſure unfit for food. If ſown much later, it does not apple, and there is no food but from the leaves.

Though by a drill-plough the ſeed may be ſown of any thickneſs, the ſafeſt way is to ſow thick. Thin-ſowing is liable to may accidents, which are far from being counterbalanced by the expence that is ſaved in thinning. Thick-ſowing can bear the ravage of the black fly, and leave a ſufficient crop behind. It is a protection againſt drought, gives the plants a rapid progreſs, and eſtabliſhes

eſtabliſhes them in the ground before it is neceſſary to thin them.

The ſowing turnip broadcaſt is univerſal in England, and common in Scotland, though a barbarous practice. The eminent advantage of turnip is, that, beſide a profitable crop, it makes a moſt complete fallow; and the latter cannot be obtained but by horſe-hoeing. Upon that account, I recommend with confidence the ſowing turnip in rows at three feet diſtance: wider rows anſwer no profitable end, ſtraiter rows afford not room for a horſe to walk in. When the turnip is about four inches high, annual weeds will appear. Go round every interval with the ſlighteſt furrow poſſible, at the diſtance of two inches from each row, moving the earth from the rows toward the middle of the interval. A thin plate of iron muſt be fixed on the left ſide of the plough, to prevent the earth from falling back, and burying the turnip. Next, let women be employed to weed the rows with their fingers; which is better, and cheaper done, than with the hand-hoe. The hand-hoe, beſide, is apt to diſturb the roots of the turnip that are to ſtand, and to leave them open to drought by removing the earth from them. The ſtanding turnip are to be at the diſtance of twelve inches from each other: a greater diſtance makes them ſwell too much; a leſs diſtance affords them not ſufficient room. A woman ſoon becomes expert in finger-weeding.

weeding. The following hint may be neceſſary to a learner. To ſecure the turnip that is to ſtand, let her cover it with the left hand; and with the right pull up the turnip on both ſides. After thus freeing the ſtanding turnip, ſhe may ſafely uſe both hands. Let the field remain in this ſtate, till the appearance of new annuals make a ſecond plowing neceſſary; which muſt be in the ſame furrow with the former, but a little deeper. As in this plowing the iron plate is to be removed, part of the looſe earth will fall back on the roots of the plants: the reſt will fill the middle of the interval, and bury every weed. When weeds begin again to appear, then is the time for a third plowing in an oppoſite direction, which lays the earth to the roots of the plants. This plowing may be about the middle of Auguſt, after which, weeds riſe very faintly. If they do riſe, another plowing will clear the ground of them. Weeds that at this time riſe in the row, may be cleared with a hand-hoe, which can do little miſchief among plants diſtant twelve inches from each other. I am certain however from experience, that it may be done cheaper with the hand *. And after that the leaves of turnips in a row

* Children under thirteen may be employed to weed turnip with the fingers I have ſeen them go on in that work with alacrity; and a ſmall premium will have a good effect.

row meet together, the hand is the only inſtrument that can be applied for weeding.

Tull was the father of horſe-hoeing huſbandry; and to deſert his method requires an apology. The apology I make is, that the method I preſcribe, and have long practiſed, is more ſimple, leſs operoſe, and, as far as I can judge, equally ſucceſsful. Our farmers ought to be excited by every motive to embrace the cultivation of turnip; and nothing will contribute more than to render it eaſy and ſimple. Tull's method is not a little intricate; and to its intricacy, I am perſuaded, is owing the neglect of it in England.

In Yorkſhire the lambs are fed in July in a turnip field. They eat the weeds without touching the young turnip. Why ſhould we clean a field for turnip, they will ſay, when it deprives the lambs of their food?

Where land is clean and the ſoil well pulverized, a crop of turnip may be procured with very little trouble. The field being laid flat without ridges, cover it with well rotted dung. With a plough having a double mouldboard, make furrows having intervals of three feet, which will mix the dung with the ſoil, and make three feet ridges; on the crowns of which drill the ſeed. As ſoon as weeds begin to appear, a deeper furrow with the ſame plough will bury the young weeds. Thin the turnip in the rows when two inches high, and gather clean earth to their roots.

If the land was perfectly clean, this will be sufficient for a good crop.

In swampy ground, the surface of which is best reduced by paring and burning, the seed may be sown in rows with intervals of a foot. To save time, a drill-plough may be used that sows three or four rows at once. Hand-hoeing is proper for such ground; because the soil under the burnt *stratum* is commonly full of roots, which digest and rot better under ground than when brought to the surface by the plough. In the mean time, while these are digesting, the ashes will secure a good crop.

The black fly is a great enemy to turnip; and I may justly say the only enemy, because every other article can be secured, by skill and diligence in the husbandman. In a rich and warm soil, where the black fly is instinctively directed to deposit its eggs, I take it to be an unconquerable enemy. Our only security is, to change the field of battle, by raising turnip in a moorish ground, or in ground newly broken up, according to the directions given above *.

2. POTATOES.

THE choice of soil is not of greater importance in any other plant than in a potato. This plant in clay soil, or in rank black loam lying low

* Chap. 4. § 2.

low without ventilation, never makes palatable food. In a gravelly or ſandy ſoil, expoſed to the ſun and to free air, it thrives to perfection, and has a good reliſh. But a rank black loam, tho' improper to raiſe potatoes for the table, produces them in great plenty; and the product is good and wholeſome food for horned cattle, hogs and poultry.

The ſpade is a proper inſtrument for raiſing a ſmall quantity, or for preparing corners or other places inacceſſible to the plough; but for raiſing potatoes in quantities, the plough is the only inſtrument.

As two great advantages of a drilled crop, are, to deſtroy weeds, and to have a fallow at the ſame time with the crop, no judicious farmer will think of raiſing potatoes in any other way. In September or October, as ſoon as that year's crop is removed, let the field have a rouſing furrow, a croſs-brakeing next, and then cleared of weeds by the cleaning harrow. Form it into three-feet ridges, in that ſtate to lie till April, which is the proper time for planting potatoes. Croſs-brake it to raiſe the furrows a little. Then lay well-digeſted horſe-dung along the furrows, upon which lay the roots at eight inches diſtance. Cover up theſe roots with the plough, going once round every row. This makes a warm bed for the potatoes; hot dung below, and a looſe covering above, that admits every ray of the ſun.

ſun. As ſoon as the plants appear above ground, go round every row a ſecond time with the plough, which will lay upon the plants an additional inch or two of mould, and at the ſame time bury all the annuals; and this will complete the plowing of the ridges. When the potatoes are ſix inches high, the plough, with the deepeſt furrow, muſt go twice along the middle of each interval in oppoſite directions, laying earth firſt to one row, and next to the other. And to perform this work, a plough with a double mouldboard will be more expeditious. But as the earth cannot be laid cloſe to the roots by the plough, the ſpade muſt ſucceed, with which four inches of the plants muſt be covered, leaving little more but the tops above ground; and this operation will at the ſame time bury all the weeds that have ſprung ſince the former plowing. What weeds ariſe after, muſt be pulled up with the hand. A hoe is never to be uſed here: it cannot go ſo deep as to deſtroy the weeds without cutting the fibres of the plants; and if it ſkim the ſurface, it only cuts off the heads of the weeds, and does not prevent their puſhing again.

The ſhorteſt and moſt perfect method of taking up potatoes, is to plow once round every row at the diſtance of four inches, removing the earth from the plants, and gathering up with the hand all the potatoes that appear. The diſtance is made four inches to prevent cutting the roots, which

which are ſeldom found above that diſtance from the row on each ſide. When the ground is thus cleared by the plough, raiſe the potatoes with a fork having three broad toes or claws, which is better than a ſpade, as it does not cut the potatoes. The potatoes thus laid above ground, muſt be gathered with the hand. By this method ſcarce a potato will be left.

As potatoes are a comfortable food for the low people, it is of importance to have them all the year round. For a long time, potatoes in Scotland were confined to the kitchen-garden; and after they were planted in the field, it was not imagined at firſt that they could be uſed after the month of December. Of late years they have been found to anſwer even till April; which has proved a great ſupport to many a poor family, as they are eaſily cooked, and require neither kiln nor mill. But there is no cauſe for ſtopping there. It is eaſy to preſerve them till the next crop: When taken out of the ground, lay in the corner of a barn a quantity that may ſerve till April, covered from froſt with dry ſtraw preſſed down: bury the remainder in a hole dug in dry ground, mixed with the huſks of dried oats, ſand, or the dry leaves of trees, over which build a ſtack of hay or corn. When the pit is opened for taking out the potatoes, the eyes of what have a tendency to puſh, muſt be cut out; and this cargo will ſerve all the month of June. To be be ſtill more certain of making the old crop meet the new, the ſetting a ſmall quantity may be delayed till June, to be taken up at the ordinary time before froſt. This cargo, having not arrived to full growth, will not be ſo ready to puſh as what are ſet in April.

If the crop happen to be exhauſted before the new crop is ready, the interval may be ſupplied by the potatoes of the new crop that lie next the ſurface, to be picked up with the hand; which, far from hurting the crop, will rather improve it.

3. Carrot and Parsnip.

Of all roots a carrot requires the deepeſt ſoil. It ought at leaſt to be a foot deep, all equally good from top to bottom. If ſuch a ſoil be not in the farm, it may be made artificially by trench-plowing, which brings to the ſurface what never had any communication with the ſun or air. When this new ſoil is ſufficiently improved by a crop or two with dung, it is fit for bearing carrots. Beware of dunging the year when the carrots are ſown; for with freſh dung they ſeldom are free of ſcabs.

Loam and ſand are the only ſoils proper for that root.

The ground muſt be prepared with the deepeſt furrow that can be taken, the ſooner after harveſt the better; immediately upon the back of which

which a ribbing ought to ſucceed as directed for barley. At the end of March or beginning of April, which is the time of ſowing the ſeed, the ground muſt be ſmoothed with a brake. Sow the ſeed in drills, with intervals of a foot for hand-hoeing; which is no expenſive operation where the crop is confined to an acre or two: but if the quantity of ground be greater, the intervals ought to be three feet, in order for horſe-hoeing.

In flat ground without ridges, it may be proper to make parallel furrows with the plough, ten feet from each other, in order to carry off any redundant moiſture.

At Parlington in Yorkſhire, from the end of September to the firſt of May, twenty work-horſes, four bullocks, and ſix milk-cows, were fed on the carrots that grew on three acres; and theſe animals never taſted any other food but a little hay. The milk was excellent: and, over and above, thirty hogs were fattened upon what was left by the other beaſts. I have this fact from undoubted authority.

The culture of parſnips is the ſame with that of carrots.

SECT.

SECT. III.

Plants cultivated for Leaves.

THERE are many garden-plants of this kind. The plants proper for the field are cabbage, red and white; colewort, plain and curled. I know very little difference in the cultivation of theſe plants. And therefore, to ſave trouble, I ſhall confine myſelf to cabbage.

Cabbage is an intereſting article in huſbandry, every ſoil being more or leſs proper for it. It is eaſily raiſed, is ſubject to few diſeaſes, reſiſts froſt more than turnip, is palatable to cattle, and ſooner fills the ſtomach than turnip, carrot, or potatoes.

The ſeaſon for ſetting cabbage, depends on the uſe it is intended for. If intended for feeding in November, December, and January, the plants muſt be ſet in March or April, from ſeed ſown the end of July the preceding year. If intended for feeding in March, April, and May, the plants muſt be ſet the firſt week of the preceding July, from ſeed ſown end of February or beginning of March the ſame year. The late ſetting of the plants retards their growth; by which means they have a vigorous growth the following ſpring. And this crop makes an important link in the chain that connects winter and ſummer green food.

food. Let me obſerve by the by, that where cabbage for ſpring-food happens to be neglected, a few acres of rye ſown at Michaelmas will ſupply the want. After the rye is conſumed, there is time ſufficient to prepare the ground for turnip.

And now to prepare a field for cabbage. Where the plants are to be ſet in March, the field muſt be made up, after harveſt, in ridges three feet wide. In that form let it lie all winter, to be mellowed with air and froſt. In March, take the firſt opportunity between wet and dry, to lay dung in the furrows. Cover the dung with a plough, which will convert the furrow into a crown, and conſequently the crown into a furrow. Set the plants upon the dung, diſtant from each other three feet. Plant them ſo as to make a ſtraight line croſs the ridges, as well as along the furrows, to which a gardener's line ſtretched perpendicularly croſs the furrows will be requiſite. This will ſet each plant at the diſtance preciſely of three feet from the plants that ſurround it. The purpoſe of this accuracy, is to give opportunity for plowing, not only along the ridges, but croſs them. This mode is attended with three ſignal advantages: it ſaves hand-hoeing, it is a more complete dreſſing to the ſoil, and it lays earth neatly round every plant.

If the ſoil be deep and compoſed of good earth, a trench-plowing after the preceding crop will not be amiſs; in which caſe, the time for dividing the field into three-feet ridges as above, ought to be immediately before the dunging for the plants.

If weeds happen to riſe ſo cloſe to the plants as not to be reached by the plough, it will require very little labour to deſtroy them with a hand-hoe.

Unleſs the ſoil be much infeſted with annuals, twice plowing after the plants are ſet will be a ſufficient dreſſing. The firſt removes the earth from the plants; the next, at the diſtance of a month or ſo, lays it back.

Where the plants are to be ſet in July, the field muſt be ribbed as directed for barley. It ought to have a ſlight plowing in June before the planting, in order to looſen the ſoil, but not ſo as to bury the ſurface-earth; after which the three-feet ridges muſt be formed, and the other particulars carried on as directed above with reſpect to plants that are to be ſet in March.

CHAP. VI.

CULTURE OF GRASS.

THE graſſes commonly ſown for paſture, for hay, or to cut green for cattle, are red clover, white clover, yellow clover, ryegraſs, and narrow-

narrow-leaved plantain, commonly called *ribwort*. Among the great variety of graſſes, there is little doubt, but that ſome new ſorts may be introduced to advantage. But without meaning to ſhut out light, I am certain that the graſſes mentioned anſwer completely all uſeful purpoſes of huſbandry.

Red clover is of all the moſt proper to be cut green for ſummer-food. It is a biennial plant when ſuffered to perfect its ſeed; but when cut green, it will laſt three years, and in a dry ſoil longer. At the ſame time, the ſafeſt courſe is to let it ſtand but a ſingle year: if the ſecond year's crop happen to be ſcanty, it proves, like a bad crop of peaſe, a great encourager of weeds by the ſhelter it affords them. Sainfoin and lucern make excellent green food; and, when preſerved clean from weeds, they will ſtand good ten or twelve years, eſpecially in a deep gravel. There they extend their roots very deep, and will grow vigorouſly in a dry ſummer, when other plants of ſhorter roots languiſh by lack of moiſture. But except in ſuch a ſoil, I venture to declare in favour of red clover; and my reaſon is, that the expence of ſowing it yearly is much leſs than that of hand-hoeing ſainfoin and lucern frequently every year. Sainfoin indeed ſeems to be the hardier plant. Farmers about Bath ſow it in their higheſt unſheltered grounds; and it endures froſt better than clover, or any ſort of grain. Like clover

clover it is ſown with barley; but as the ſeed is ſmall, much leſs will ſuffice than of barley. It is commonly cut but once the firſt year; and the firſt winter, coal-aſhes are ſpread upon it.

Here, as in all other crops, the goodneſs of red-clover ſeed is of importance. Chuſe plump ſeed of a purple colour, becauſe it takes on that colour when ripe. It is red when hurt in the drying, and of a faint colour when unripe.

Red clover is luxuriant upon a rich ſoil, whether clay, loam, or gravel; it will grow even upon a moor properly cultivated. A wet ſoil is its only bane; for there it does not thrive.

To have red clover in perfection, weeds muſt be extirpated, and ſtones taken off. The mould ought to be made as fine as harrowing can make it; and the ſurface be ſmoothed with a light roller, if not ſufficiently ſmooth without it. This gives opportunity for diſtributing the ſeed evenly; which muſt be covered by a ſmall harrow with teeth no larger than of a garden-rake, three inches long, and ſix inches aſunder; of which there is a draught annexed. In harrowing, the man ſhould walk behind with a rope in his hand fixed to the back part of the harrow, ready to diſentangle it from ſtones, clods, turnip or cabbage roots, which would trail the ſeed, and diſplace it.

Nature has not determined any preciſe depth for the ſeed of red clover, more than of other ſeed. It will grow vigorouſly from two inches deep, and it it will grow when barely covered. Half an inch I reckon the moſt advantageous poſition in clay ſoil, a whole inch in what is light or looſe. It is a vulgar error, that ſmall ſeed ought to be ſparingly covered. Miſled by that error, farmers commonly cover their clover-ſeed with a thorn buſh; which not only covers it unequally, but leaves part on the ſurface to wither in the air.

The proper ſeaſon for ſowing red clover, is from the middle of April to the middle of May. It will ſpring from the 1ſt of March to the end of Auguſt; but ſuch liberty ought not to be taken except from neceſſity.

The ordinary manner of ſowing, is to hold the ſeed between the thumb and the two fore fingers. I prefer another manner; which is, to take into the hollow of the hand as much as will ſerve in walking three or four ſteps; and to join the fingers to the palm ſo looſely, as to make many ſmall holes for throwing out the ſeed. A hand cloſed ſo looſely, reſembles the roſe of a watering pan, which ſcatters the water through a number of ſmall holes.

There is not a greater miſtake in huſbandry, than to be ſparing of ſeed. Ideal writers talk of ſowing an acre with four pounds. That quantity of ſeed, ſay they, will fill an acre with plants as thick as they ought to ſtand. I admit this rule to be good where grain is the object; but not with reſpect to graſs. Graſs-ſeed cannot be ſown

too

too thick: the plants ſhelter one another: they retain the dew: and they muſt puſh upward, having no room laterally. Obſerve the place where a ſack of peaſe, or of other grain, has been ſet down for ſowing: the ſeed dropt there accidentally grows more quickly than in the reſt of the field ſown thin out of hand: I have ſeen it ſix inches high, when the reſt of the ſeed ſcarce appeared above ground. A young plant of clover, or of ſainfoin, according to Tull, may be raiſed to a great ſize where it has room; but the field will not produce half the quantity. When red clover is ſown for cutting green, there ought not to be leſs than twenty-four pounds to an acre. A field of clover is ſeldom too thick: the ſmaller a ſtem be, the more palatable to cattle. When too thin, the ſtems tend to wood.

Red clover is commonly ſown with grain; and the doubt may be, what grain is the moſt proper. I pronounce in favour of flax; and I pronounce with confidence from multiplied experience. The ſoil muſt be highly cultivated for flax as well as for red clover. The proper ſeaſon of ſowing is the ſame for both: the leaves of flax being very ſmall, admit of free circulation of air; and flax being an early crop, is removed ſo early as to give the clover time for growing. In a rich ſoil it has grown ſo faſt, as to afford a good cutting that very year. Next to flax, barley is the beſt companion to clover. The ſoil muſt be looſe and free

for

for barley ; and ſo it ought to be for clover : the ſeaſon of ſowing is the ſame ; and the clover is well eſtabliſhed in the ground, before it is overtopped by the barley. At the ſame time, barley commonly is ſooner cut than either oats or wheat. In a word, barley is rather a nurſe than a ſtepmother to clover during its infancy. When clover is ſown in ſpring upon wheat, the ſoil, which has lain five or ſix months without being ſtirred, is an improper bed for it ; and the wheat, being in the vigour of growth, overtops it from the beginning. It cannot be ſown along with oats, becauſe of the hazard of froſt ; and when ſown as uſual among oats three inches high, it is overtopped, and never enjoys free air till the oats are cut. Add, that where oats are ſown upon the winter-furrow, the ſoil is rendered as hard as when under wheat. Red clover is ſometimes ſown by itſelf, without other grain ; but this method, beſide loſing a crop, is not ſalutary ; becauſe clover in its infant ſtate requires ſhelter. The year 1775 confirmed all my experiments about clover. In part of a field, flax was ſown with it ; in another part, barley ; and, in the remainder, it was ſown alone. The clover on the firſt was the beſt ; on the ſecond, inferior ; and, on the third, the worſt. Yet the barley was ſown thin, as it ought to be with graſs-ſeeds, and was but an indifferent crop.

And this leads to the quantity of grain proper to be ſown with clover. In a rich ſoil well pulverized,

verized, a peck of barley on an Engliſh acre is all that ought to be ventured; but there is not much ſoil in Scotland ſo rich. Two Linlithgow firlots make the proper quantity for an acre that produces commonly ſix bolls of barley; half a firlot for what produces nine bolls. To thoſe who are governed by cuſtom, ſo ſmall a quantity will be thought ridiculous. Let them only conſider, that a rich ſoil in perfect good order, will from a ſingle ſeed of barley produce twenty or thirty vigorous ſtems. People may flatter themſelves with the remedy, of cutting barley green for food, if it happen to oppreſs the clover. This is an excellent remedy in a field of an acre or two; but the cutting an extenſive field for food muſt be ſlow; and while one part is cutting, the clover is ſmothered in other parts.

The culture of white clover, of yellow clover, of ribwort, of ryegraſs, is the ſame in general with that of red clover. I proceed to their peculiarities. Yellow clover, ribwort, ryegraſs, are all of them early plants, blooming the end of April or beginning of May. The two latter are evergreens, and therefore excellent for winter-paſture. Ryegraſs is leſs hurt by froſt than any of the clovers, and will thrive in a moiſter ſoil: nor in that ſoil is it much affected by drought. In a rich ſoil, it grows four feet high: even in the dry ſummer 1775, it roſe to three feet eight inches; but it had gained that height before the drought

drought came on. Theſe graſſes are generally ſown with red clover for producing a plentiful crop. The proportion of ſeed is arbitrary; and there is little danger of too much. When ryegraſs is ſown for procuring ſeed, five firlots wheat-meaſure may be ſown on an acre; and for procuring ſeed of ribwort, forty pounds may be ſown. The roots of ryegraſs ſpread horizontally: they bind the ſoil by their number; and though ſmall, are yet ſo vigorous as to thrive in hard ſoil. Red clover has a large tap-root, which cannot penetrate any ſoil but what is open and free; and the largeneſs of the root makes the ſoil ſtill more open and free. Ryegraſs, once a great favourite, appears to be diſcarded in moſt parts of Britain. But were the management of it well underſtood, it would be reſtored to high favour. The common practice has been, to ſow it with red clover, and to cut them promiſcuouſly the beginning of June for green food, and a little later for hay. This indeed is the proper ſeaſon for cutting red clover, becauſe at that time it begins to flower; but as at that time the ſeed of the ryegraſs is approaching to maturity, its growth is at an end for that year, as much as of oats or barley cut after the ſeed is ripe. Oats or barley cut green before the ſeed forms, will afford two other cuttings; which alſo is the caſe of ryegraſs, of yellow clover, and of ribwort. By ſuch management, all the profit will be drawn that theſe plants can afford.

When red clover is intended for ſeed, the ground ought to be cleared of weeds, were it for no other purpoſe than that the ſeed cannot otherwiſe be preſerved pure: what weeds eſcape the plough, ought to be taken out by the hand. In England, when a crop of ſeed is intended, the clover is always firſt cut for hay. This I conjecture to be done, as in fruit-trees, to check the growth of the wood, in order to encourage the fruit. This practice will not anſwer with us, as the ſeed would often be too late of ripening. Better to eat the clover with ſheep till the middle of May, which will give time for the ſeed to ripen. The firſt crop of red clover, if not retarded by eating, will not anſwer ſo well for ſeed as the ſecond. The plants grow vigorouſly; and the leaves cover the ſeed ſo as to give little acceſs to the ſun and air. For the ſame reaſon, peaſe on rich land grow to the ſtraw, and ſeldom produce much ſeed. The ſeed is ripe when, upon rubbing it between the hands, it parts readily from the huſk. Then apply the ſcythe, ſpread the crop thin, and turn it carefully. When perfectly dry, take the firſt opportunity of a hot day for threſhing it on boards covered with a coarſe ſheet. Another way leſs ſubject to riſk, is to ſtack the dry hay, and to threſh it the end of April. After the firſt threſhing, expoſe the huſks to the ſun, and threſh then them over and over till no ſeed remain. Nothing is more efficatious than a hot ſun to make the huſk part with its ſeed; in which view, it may be expoſed to the ſun by parcels, an hour or two before the flail is applied.

White clover intended for ſeed, is managed in the ſame manner. No plant ought to be mixed with ryegraſs that is intended for ſeed. In Scotland, much ryegraſs-ſeed is hurt by tranſgreſſing that rule. The ſeed is ripe when it parts eaſily from the huſk. The yellowneſs of the ſtem is another indication of its ripeneſs; in which particular it reſembles oats, barley, and other culmiferous plants. The beſt manner to manage a crop of ryegraſs for ſeed, is to bind it looſely in ſmall ſheaves, widening them at the bottom to make them ſtand erect; as is done with oats in moiſt weather. In that ſtate, they may ſtand till ſufficiently dry for threſhing. By this method, they dry more quickly, and are leſs hurt by rain, than by cloſe binding and putting the ſheaves in ſhocks like corn. The worſt way of all, is, to ſpread the ryegraſs on the moiſt ground; for it makes the ſeed malten. The ſheaves, when ſufficiently dry, are carried in cloſe carts to where they are to be threſhed on a board, as mentioned above for clover. Put the ſtraw in a rick, when a hundred ſtone or ſo are threſhed. Carry the threſhing-board to the place where another rick is intended; and ſo on till the whole ſeed is threſhed,

and

and the ſtraw ricked. There is neceſſity for cloſe carts to ſave the ſeed, which is apt to drop out in a hot ſun: and, as obſerved above, a hot ſun ought always to be choſen for threſhing. Carry the ſeed in ſacks to the granary or barn, there to be ſeparated from the huſks by a fanner. Spread the ſeed thin upon a timber floor, and turn it once or twice a-day, till perfectly dry. If ſuffered to take a heat, it is uſeleſs for ſeed.

I ſhall conculde this chapter with a few obſervations upon the different endurance of plants. That ſome plants muſt be perpetual is evident from the old graſs fields ſeen in every quarter, that are always paſtured, and never ſuffered to go to ſeed. Even the largeſt trees have a period of exiſtence; but ſuch graſs plants can never wear out, otherways the field would be left bare. Some plants endure but a year; wheat, for example, oats, barley, peaſe. Their deſtined purpoſe is to carry ſeed, which is compleated within the year, and then they die. Some plants require two years to perfect their ſeed, which is the caſe of cabbage. Red clover, yellow clover, ryegraſs, carry ſeed every year; but the firſt ſeldom lives above two years, and the others not above ſeven or eight. White clover is a perenniel plant. Like a ſtrawberry it throws out flagellæ or runners from the ſtem, which take root and become new plants without end. From the roots of others there are ſtems produced yearly, which do not

not perfect their seed till the second year; after which these stems die and give place to others. Such plants have at the same time not only stems of the second year bearing seed, but recent stems that require another year's growth to carry seed. It is of perenniel plants that the green covering of an old pasture field is composed. Some of these are more palatable and nourishing than others; and the old grass field is valuable in proportion to the goodness of plants that grow in it.

But to give all the satisfaction possible on a subject of capital importance in husbandry, I here annex a list of perenniel grasses, including not only what endure for ever, but what are very long lived. Every pasture field in Scotland from 5 s. to L. 5 *per* acre, is composed of some of the sixteen following grasses.

1. *Poa annua*. Suffolk grass.

This grass requires a mild climate more than any other gramineous plant we have. It perishes by the severity of the Swedish winter; and upon that account is classed among annuals by Linnæus. But though it is an annual in Sweden, it is perenniel in Scotland.

2. *Poa trivialis*. Common meadow-grass.
3. *Poa pratensis*. Great meadow-grass.
4. *Poa angustifolia*. Narrow-leafed meadow-grass.

5. *Lolium*

5. *Lolium perenne.* Rye-graſs.
6. *Avena flaveſcens.* Yellow oat-graſs.

The moſt valuable paſtures both in England and Scotland, are moſtly formed of theſe ſix ſpecies, which never grow but in rich ſoil. They affect old manured land, and prevail in our croft lands; but are never to be ſeen in what we call outfield, nor will they grow in it.

7. *Alopecurus pratenſis.* Meadow fox tail.
8. *Anthoxanthum odoratum.* Vernal graſs.
9. *Cynoſurus criſtatus.* Creſted dogs tail.
10. *Agroſtis cappilaris.* Fine bent graſs.

Theſe four ſpecies, though often mixed with the former in rich paſture, can take up with a meaner ſoil. They compoſe the chief part of all the outfield paſture, that is, of all the paſture in Scotland. The tenth ſpecies is the moſt prevalent; and forms more of the herbage of Scotland than any other plant.

11. *Phleum pratenſe.* Meadow cats tail.
12. *Holcus lanatus.* Meadow ſoft-graſs.
13. *Dactylis glomerata.* Orchard-graſs.
14. *Bromus ſecalinus.* Gooſe-graſs.

Theſe four are tall rank graſſes, that are poſtponed by cattle when green, where they can get any of the former. They afford good hay however, and give a great crop.

15. *Feſtuca*

15. *Festuca ovina.* Sheeps fescue.
16. *Festuca duriuscula.* Hard fescue.

These two species form our best sheep pasture; but are of too diminitive a growth for the scythe, and not a subject of culture.

All these grasses propagate powerfully by the roots as well as by seed. Their manner of growth by the roots, as above hinted, is in two different ways. The first by suckers, which rise successively from the present plant; and the grasses that grow in this way, form large tufts of a round figure. The second is by runners, which shoot forth on all sides of the plant. These are the grasses which form the most close uniform sward, and the best pasture. Both kinds are perenniel; and I may say perpetual; for in their proper soil they will grow for ever, without being renewed by seed.

Ribwort and white clover, are not in this list; because they are not considered as grasses by the learned in botany. But as they are perenniel, propagating both by seeds and roots, they are capital articles in old pasture. Ryegrass in its natural soil is a perenniel: in that soil, it propagates both by seeds and roots, and subsists in vigour where it never was sown by art. In a soil less natural to it, the seeds sown may produce a few good crops; but it will not renovate itself either by seeds or roots. This is the case in most fields

fields of cultivated ryegraſs, eſpecially in a wet or clay ſoil, or where other predominant graſſes prevail. Even white clover will decay, where it is raiſed in a ſoil that is not natural to it.

CHAP. VII.

Rotation of Crops.

NO branch of huſbandry requires more ſkill and ſagacity than a proper rotation of crops, ſo as to keep the ground always in heart, and yet to draw out of it the greateſt profit poſſible. A horſe is purchaſed for labour ; and it is the purchaſer's intention to make the moſt of him. He is well fed, and wrought according to his ſtrength : to overwork him, is to render him uſeleſs. Preciſely ſimilar is land. Profit is the farmer's object ; but he knows, that to run out his farm by indiſcreet cropping, is not the way to make profit. Some plants rob the ſoil, others are gentle to it : ſome bind, others looſen. The nice point is, to intermix crops, ſo as to make the greateſt profit conſiſtently with keeping the ſoil in order. In that view, the nature of the plants employed in huſbandry, muſt be accurately examined.

The difference between culmiferous and leguminous plants, is occaſionally mentioned above *.
With

* Chap. 5.

With reſpect to the preſent ſubject, a narrower inſpection is neceſſary. Culmiferous plants, having ſmall leaves and few in number, depend moſtly on the ſoil for nouriſhment, and little on the air. During the ripening of the ſeed, they draw probably their whole nouriſhment from the ſoil; as the leaves by this time, being dry and withered, muſt have loſt their power of drawing nouriſhment from the air. Now, as culmiferous plants are chiefly cultivated for ſeed, and are not cut down till the ſeed be fully ripe, they may be pronounced all of them to be robbers, ſome more ſome leſs. But ſuch plants, while young, are all leaves; and in that ſtate draw moſt of their nouriſhment from the air. Hence it is, that where cut green for food to cattle, a culmiferous crop is far from being a robber. A hay-crop accordingly, even where it conſiſts moſtly of ryegraſs, is not a robber, provided it be cut before the ſeed is formed; which at any rate it ought to be, if one will have hay in perfection. And the foggage, excluding froſt by covering the ground, keeps the roots warm. A leguminous plant, by its broad leaves, draws much of its nouriſhment from the air. A cabbage, which has very broad leaves and a multitude of them, owes its growth more to the air than to the ſoil. One fact is certain, that a cabbage cut and hung up in a damp place, preſerves its verdure longer than other plants. At the ſame time, a ſeed is that part of a plant which requires the moſt nouriſhment; and for that nouriſhment a culmiferous plant muſt be indebted entirely to the ſoil. A leguminous crop, on the contrary, when cut green for food, muſt be very gentle to the ground. Peaſe and beans are leguminous plants; but being cultivated for ſeed, they ſeem to occupy a middle ſtation: their ſeed makes them more ſevere than other leguminous crops cut green: their leaves, which grow till reaping, make them leſs ſevere than a culmiferous plant left to ripen.

Theſe plants are diſtinguiſhed no leſs remarkably by the following circumſtance. All the ſeeds of a culmiferous plant ripen at the ſame time. As ſoon as they begin to form, the plant becomes ſtationary, the leaves wither, the roots ceaſe to puſh, and the plant, when cut down, is blanched and ſapleſs. The ſeeds of a leguminous plant are formed ſucceſſively: flowers and fruit appear at the ſame time in different parts of the plant. This plant accordingly is continually growing, and puſhing its roots. Hence the value of bean or peaſe ſtraw above that of wheat or oats: the latter is withered and dry when the crop is cut; the former, green and ſucculent. The difference therefore with reſpect to the ſoil between a culmiferous and leguminous crop, is great. The latter, growing till cut down, keeps the ground in conſtant motion, and leaves it to the plough looſe and mellow. The former gives

over

over growing long before reaping, and the ground, by want of motion, turns compact and hard. Nor is this all. Dew falling on a culmiferous crop after the ground begins to harden, rests on the surface, and is sucked up by the next sun. Dew that falls on a leguminous crop, is shaded from the sun by the broad leaves, and sinks at leisure into the ground. The ground accordingly after a culmiferous crop, is not only hard but dry: after a leguminous crop, it is not only loose, but soft and unctuous.

Of all culmiferous plants, wheat is the most severe, by the long time it occupies the ground without admitting a plough. And as the grain is heavier than that of barley or oats, it probably requires more nourishment than either. Spring-wheat is creeping into use: if it succeed, it will probably be not much more severe than other culmiferous plants. It is observed above, that as pease and beans draw part of their nourishment from the air by their green leaves while allowed to stand, they draw the less from the ground; and by their constant growing they leave it in good condition for subsequent crops. In both respects they are preferable to any culmiferous crop.

Culmiferous crops, as observed above, are not robbers when cut green: the soil, far from hardening, is kept in constant motion by the pushing

ing of the roots, and is more tender than if it had been left at reſt without bearing any crop.

Bulbous-rooted plants are above all operative in dividing and pulveriſing the ſoil. Potato-roots grow ſix, eight, or ten, inches under the ſurface; and, by their ſize and number, they divide and pulverize the ſoil better than can be done by the plough; conſequently, whatever be the natural colour of the ſoil, it is black when a potato-crop is taken up. The potato, however, with reſpect to its quality of dividing the ſoil, muſt yield to a carrot or parſnip; which are large roots, and pierce often to the depth of eighteen inches. The turnip, by its tap-root, divides the ſoil more than can be done by a fibrous-rooted plant; but as its bulbous root grows moſtly above ground, it divides the ſoil leſs than the potato, the carrot, or the parſnip. Red clover, in that reſpect, may be put in the ſame claſs with turnip.

Whether potatoes or turnip be the more gentle crop, appears a puzzling queſtion. The former bears ſeed, and probably draws more nouriſhment from the ſoil, than the latter when cut green. On the other hand, potatoes divide the ſoil more than turnip, and leave it more looſe and friable. It appears no leſs puzzling, to determine between cabbage and turnip: the former draws more of its nouriſhment from the air, the latter leaves the ſoil more free and open.

Here

Here are a number of facts: What is the reſult of the whole? Clearly what follows. Culmiferous plants are robbers; ſome more, ſome leſs: they at the ſame time bind the ſoil; ſome more, ſome leſs. Leguminous plants in both reſpects are oppoſite: if any of them rob the ſoil, it is in a very ſlight degree; and all of them without exception looſen the ſoil. A culmiferous crop, however, is generally the more profitable: but few ſoils can long bear the burden of ſuch crops, unleſs relieved by interjected leguminous crops. Theſe, on the other hand, without a mixture of culmiferous crops, would ſoon render the ſoil too looſe.

Theſe preliminaries will carry the farmer ſome length in directing a proper rotation of crops. Where dung, lime, or other manure, can be procured in plenty to recruit the ſoil after ſevere cropping, I know no rotation more proper or profitable in a ſtrong ſoil, than wheat, peaſe or beans, barley, oats, fallow. The whole farm may be brought under this rotation, except ſo far as hay is wanted. But as ſuch command of manure is rare, it is of more importance to determine what ſhould be the rotation, where no manure can be procured but the dung collected in the farm. Conſidering that culmiferous crops are the more profitable in rich land, it would be proper to make them more frequent than the other kind. But as there are few ſoils in Scotland

land that will admit ſuch frequent culmiferous crops without ſuffering, it may be laid down as a general rule, that alternate crops, culmiferous and leguminous, ought to form the rotation. Nor are there many ſoils that will ſtand good, even with this favourable rotation, unleſs relieved from time to time by paſturing a few years. If ſuch extended rotation be artfully carried on, I take it for granted, that crops without end may be obtained in a tolerable good ſoil, without any manure but what is produced in the farm.

Having diſcuſſed the nature of plants as far as rotation of crops is concerned, the nature of the ſoil comes next under conſideration. It is ſcarce neceſſary to be mentioned, being known to every farmer, that clay anſwers beſt for wheat, moiſt clay for beans, loam for barley and peaſe, light ſoil for turnip, ſandy ſoil for rye and buck wheat; and that oats thrive better in coarſe ſoil than any other grain. Now, in directing a rotation, it is not ſufficient that a culmiferous crop be always ſucceeded by a leguminous: attention muſt be alſo given, that no crop be introduced that is unfit for the ſoil. Wheat, being a great binder, requires more than any other crop, a leguminous crop to follow. But every ſuch crop is not proper: potatoes are the greateſt openers of ſoil; but they are improper in a wheat-ſoil. Neither will turnip anſwer, becauſe they require a light ſoil. A very looſe ſoil, after a crop of rye requires

quires ryegrafs to bind it, or the treading of cattle in pafturing: but to bind the foil wheat muft not be ventured; for it fucceeds ill in loofe foil.

Another confideration of moment in directing the rotation, is to avoid crops that encourage weeds. Peafe is the fitteft of all crops for fucceeding to wheat, becaufe it renders the ground loofe and mellow, and the fame foil agrees with both. But beware of peafe, unlefs the foil be left by the wheat perfectly free of weeds; becaufe peafe, if not an extraordinary crop, fofter weeds. Barley may be ventured after wheat, if the farmer be unwilling to lofe a crop. It is indeed a robber; better however any crop than run the hazard of poifoning the foil with weeds. But to prevent the neceffity of barley after wheat, the land ought to be fallowed before the wheat: it cleans the ground thoroughly, and makes peafe a fecure crop after wheat. And after a good crop of peafe, barley never fails. A horfe-hoed crop of turnip is equal to a fallow for rooting out weeds; but turnip does not fuit land that is proper for wheat. Cabbage does well in wheat-foil; and a horfe-hoed crop of cabbage, which eradicates weeds, is a good preparation for fpring-wheat. A crop of beans diligently hand-hoed is in that view little inferior *. As red clover

* Spring wheat will not anfwer in ftrong clay, as it has not fufficient time to ripen.

clover requires the ground to be perfectly clean, a good crop of it ensures wheat, and next pease. In loam, a drilled crop of turnip or potatoes prepares the ground, equal to a fallow, for the same succession.

Another rule is, to avoid a frequent repetition of the same species; for to produce good crops, change of species is no less necessary than change of seed. The same species returning every second or third year, will infallibly degenerate, and be a scanty crop. This is remarkably the case of red clover. Nor will our fields bear pleasantly perpetual crops of wheat after fallow, which is the practice of some English farmers.

Hitherto of rotation in the same field. I add one rule concerning rotation in different fields; which is, to avoid crouding crops one after another in point of time; but to chuse such as admit intervals sufficient for leisurely dressing, which gives opportunity to manage all with the same hands, and with the same cattle; for example, beans in January or February, pease and oats in March, barley and potatoes in April, turnip in June or July, wheat and rye in October.

For illustrating the foregoing rules, a few instances of exceptionable rotations, will not be thought amiss. The following is an usual rotation in Norfolk. First, wheat after red clover. Second, barley. Third, turnip. Fourth, barley with red clover. Fifth, clover cut for hay. Sixth, a second

ſecond year's crop of clover commonly paſtured. Dung is given to the wheat and turnip. Againſt this rotation ſeveral objections lie. Barley after wheat is improper. The two crops of barley are too near together. The ſecond crop of clover muſt be very bad, if paſturing be the beſt way of conſuming it; and if bad, it is a great encourager of weeds. But the ſtrongeſt objection is, that red clover repeated ſo frequently in the ſame field cannot fail to degenerate; and of this the Norfolk farmers begin to be ſenſible. Salton in Eaſt Lothian is a clay ſoil; and the rotation there is, wheat after fallow and dung. Second, barley after two plowings; the one before winter, the other immediately before the ſeed is ſown. Third, oats. Fourth, peaſe. Fifth, barley. Sixth, oats: and then fallow. This rotation conſiſts chiefly of robbing crops. Peaſe are the only leguminous crop, which even with the fallow is not ſufficient to looſen a ſtiff ſoil. But the ſoil is good, which in ſome meaſure hides the badneſs of the rotation. About Seaton, and all the way from Preſton to Goſsford, the ground is ſtill more ſeverly handled: wheat after fallow and dung, barley, oats, peaſe, wheat, barley, oats, and then another fallow. The ſoil is excellent; and it ought indeed to be ſo, to ſupport many rounds of ſuch cropping.

But let not our wonder be confined to this narrow ſpot; for what better do we find in the greateſt part of this county? In the pariſhes of Tranent, Aberlady, Dirleton, Northberwick, and Athelſtonefoord, the following rotations were formerly univerſal, and to this day are much more frequent than any other mode.

1. After fallow with dung, wheat, barley, oats, peaſe and beans, barley, oats, wheat.

2. After fallow and dung, barley, oats, peaſe and beans, wheat, barley, oats, peaſe, wheat.

3. After fallow and dung, wheat, oats, peaſe, barley, oats, wheat.

4. After fallow and dung, barley, oats, beans, wheat, peaſe, barley, oats.

Eaſt Lothian, time out of mind, has been famous for ſuperior ſkill in agriculture; and yet, to ſeek for inſtruction there, one would be greatly miſled. That county, for the richneſs of its crops, is more indebted to the fertility of its ſoil, than to the ſkill of its farmers. What pity it is, that ſo fine a country ſhould be poſſeſſed by men ſo little grateful to nature for her bounties! But their ingratitude is not ſuffered to paſs with impunity. Every farmer complains that his crops are not ſuch as they have been: the decay is viſible: but the cauſe, however obvious, is not manifeſt to every one. Some few who juſtly aſcribe the decay to ſeverity of cropping, have ventured upon ſome alterations; but ſo imperfectly, from the prepoſſeſſion of former practice, as to have made no conſiderable progreſs. The following rotations, held

held to be improvements upon the former practice, will juſtify this obſervation.

1. After fallow without dung, barley, clover. Dung laid on the clover-ſtubble, and after a ſingle plowing, wheat, barley, oats.

2. After fallow with dung, wheat, barley, peaſe, wheat.

3. After fallow with dung, wheat, beans hand-hoed, wheat, peaſe and beans drilled, wheat with dung, barley, clover. Dung laid on the clover-ſtubble. Spring-wheat, one furrow before winter, and one before ſowing. Turnip broadcaſt, barley and graſs-ſeeds for paſture.

4. After fallow with dung, wheat, barley with clover-ſeed. Clover made hay two years. After the ſecond year's crop, fallow and dung for wheat, barley, oats, peaſe, oats or wheat, turnip with dung, barley, peaſe, barley, oats.

5. Potatoes dunged, wheat, peaſe, barley and graſs-ſeeds.

6. After fallow with dung, wheat, oats, peaſe or clover, wheat, oats, clover. The clover-ſtubble dunged for barley, oats, peaſe after two furrows, wheat after one furrow.

7. Lincoln barley upon ground opened from paſture graſs, peaſe, wheat, turnip, barley, clover. The ſtubble dunged for wheat, beans, barley, oats.

8. After fallow and dung, barley, clover red and white for hay, paſture ſeven years, oats, peaſe, wheat, barley, oats.

9. Turnip

9. Turnip after lime and dung. Barley, with which white clover, yellow clover, ryegrafs, and ribwort, are fown. Paftured with fheep feven years, broken up for oats, drilled beans, barley, oats.

Thefe rotations are far from being orthodox; but it is good to be convinced of an error: and when once a reformation is fairly begun, it is to be hoped, that the farmers in the beft county of Scotland, will at laft fettle in fuch a rotation of crops, as will prove no lefs beneficial to themfelves than to their landlords.

After fuch fevere cenfure I would gladly make fome apology for the Eaft-Lothian farmers. In Young's feveral Tours through the beft counties of England, examples are found without end of rotations no lefs exceptionable than many of thofe mentioned.

Where a field is laid down for pafture in order to be recruited, it is commonly left in that ftate many years; for it is the univerfal opinion, that the longer it lies, the richer it becomes for bearing corn. This I believe to be true; but in order to determine the mode of cropping, the important point is, what upon the whole is the moft profitable rotation; not what may produce luxuriant crops at a diftant period. Upon that point, I have no hefitation to affirm, that the farmer who keeps a field in pafture beyond a certain time, lofes every year confiderably; and that a few luxuriant

xuriant crops of corn, after twenty years of pa-ſture, and ſtill more after thirty, will not make up the loſs. The novelty of this propoſition will diſcredit it with the generality; but as the ſubject is of capital importance in the management of a farm, I muſt notwithſtanding hope for a patient hearing. Graſs-ſeeds intended for hay, produce the weightieſt crop at firſt: were hay to be taken many years ſucceſſively without manure, the crops would turn exceeding ſcanty. To prevent loſs, the farmer confines himſelf to two or three crops; and then ſurrenders the field to paſture. The ſame happens where a field is laid down with pa-ſture-graſſes: the firſt year's paſture is more plentiful than any of the ſucceeding. For a proof of which, a field newly laid down for paſture will draw a greater rent, than after being paſtured ſeven or eight years; if I ſhould ſay double, it would not be far from the truth. Nor is it difficult to aſſign a cauſe for the degeneracy. Of the plants cultivated for hay and paſture, few are long lived. Red clover, the chief of them, is only a biennial; and neither ryegraſs nor yellow clover, laſts above ſeven or eight years. They puſh vigorous ſtems at firſt, which every year turn weaker and weaker. In the mean time, the paſture is ſcanty till natural graſſes ſpring up to ſupply what are gone; which ſeldom equal ſown graſſes for feeding. The preciſe time when graſs ought to yield to corn, depends greatly on the

nature

nature of the ſoil. Our beſt director is practice. I will only venture to ſuggeſt, that clay ſoil may, without loſs, be kept in paſture much longer than gravelly or light ſoil. The former being retentive of moiſture, preſerves graſs in vigour, even during the heat of ſummer: graſs in the latter ſoon withers by lack of moiſture. This is providential: a light or gravelly ſoil can be cultivated almoſt in any ſeaſon: clay is extremely tickliſh; and where ſucceſsfully laid down for paſture, ought not raſhly to be taken up again; eſpecially as it maintains more cattle, and conſequently receives more dung, than the other kind. It is extremely true, that a greater number of corn-crops may be ſucceſsfully taken upon a field that has been long in paſture, than when paſtured but a few years. But will the additional crops of corn overbalance the mean returns from older graſs? Far from it. Why not then reſtore a field to corn as ſoon as the paſture begins to fail? The corn will quadruple the value of the paſture: the labour indeed and expence is greater; but they will be amply recompenſed by the profit.

If the leſſon here inculcated be ſolidly founded, it muſt produce a great change in the management of a farm. Paſture-graſs, while young, maintains many animals; and the field is greatly recruited by what they drop: it is even recruited by hay-crops, provided the graſs be cut before ſeeding. But the field ought to be taken up for corn

corn when the paſture begins to fail: and after a few crops, it ought to be laid down again with graſs-feeds. Seduced by a chimerical notion, that a field, by frequent corn-crops, is fatigued and requires reſt, like a labouring man or animal, careful farmers give long reſt to their fields by paſture, never adverting that it affords little profit. Pity it is, that by a chimerical notion they ſhould be tantalized, to neglect good crops within their reach. It ought to be their ſtudy to improve the ſoil, by making it free and alſo retentive of moiſture. If they accompliſh theſe ends, they need not be afraid of exhauſting the ſoil by cropping *

Againſt frequent changes from graſs to corn, what follows has the appearance of an objection, that much labour is requiſite to convert old paſture land to good tilth, reiterated plowing and harrowing, frequent brakeing, and carts after all to remove the graſs-knots. Long experience encourages me to recommend trench-plowing as the moſt effectual, cheapeſt, and moſt expeditious method for operating this converſion. With a paring-plough the ſurface is laid at the bottom of the furrow with the graſsy ſide under; and covered

* As vegetable and animal food are equally natural to man, it is admirable in Providence, to adjuſt the ſoil we tread on ſo happily to our nature, as to yield more food by a rotation of corn and graſs crops, than if it were confined to either.

covered with three or four inches of fresh mould, raised by another plough going in the same track. This being done before winter, the frost preparation makes the soil a fine bed for the seed when the season opens. Where the soil is tolerably tender, trench-plowing never fails to lay the surface smooth: the seed is all laid at an equal depth and springs up equally. Much seed also is saved, which in ordinary plowing is buried by the roughness of the surface. There is another advantage above all the rest, especially in light soil, that the moisture is retained by the grass at the bottom of the furrow, and gives great nourishment to the young plants. Old grass is generally acknowledged to be the most nourishing, to produce finer meat, and richer milk and butter, than when it was young. And it will be objected, that the farmer is deprived of that benefit by the rotation above recommended. Supposing old grass to be a benefit with respect to profit, the farmer's chief object; he pays a very high price for old grass by abstaining from the profit of such rotation. I yield, however, that where in a farm there happens to be very old grass of a good quality, the most prudent way is to let it remain as it is. And with respect to gentlemen of fortune, it may be commendable luxury to set apart for old grass a field adjacent to the mansion-house, never to be converted into corn. But the quality of the grass ought to be good; otherways the field will contribute

tribute to luxury as little as to profit. A ſoil has no choice in its plants; but foſters indifferently every kind palatable or unpalatable. In old paſture, nothing is more common than cattle every now and then putting out at the ſide of the mouth certain graſſes; an evident proof that they are unpalatable. This never is ſeen in new paſture from choice plants. Therefore, to have old paſture in perfection, let the field be ſtored with white clover, ribwort, and other ſucculent perennial plants, ſo thick ſown as to exclude all other plants. Unleſs where this precaution has been uſed, it is a great chance to find old paſture that will give abſolute ſatisfaction.

Where a farmer has acceſs to no manure but what is his own production, the caſe under conſideration, there are various rotations of crops, all of them good, though perhaps not equally ſo. I ſhall begin with two examples, one in clay, and one in free ſoil, each of the farms ninety acres. Six acres are to be incloſed for a kitchen-garden, in which there muſt be annually a crop of red clover, for ſummer-food to the working cattle. As there are annually twelve acres in hay, and twelve in paſture, a ſingle plough with good cattle will be ſufficient to command the remaining ſixty acres.

Rotation in a clay ſoil.

Incloſ.	1775.	1776.	1777.	1778.	1779.	1780.
1.	Fallow.	Wheat.	Peaſe.	Barley.	Hay.	Oats.
2.	Wheat.	Peaſe.	Barley.	Hay.	Oats.	Fallow.
3.	Peaſe.	Barley.	Hay.	Oats.	Fallow.	Wheat.
4.	Barley.	Hay.	Oats.	Fallow.	Wheat.	Peaſe.
5.	Hay.	Oats.	Fallow.	Wheat.	Peaſe.	Barley.
6.	Oats.	Fallow.	Wheat.	Peaſe.	Barley.	Hay.
7.	Paſture.	Paſture.	Paſture.	Paſture.	Paſture.	Paſture.

When the rotation is completed, the ſeventh incloſure having been ſix years in paſture, is ready to be taken up for a rotation of crops, which begins with oats in the year 1781, and proceeds as in the ſixth incloſure. In the ſame year 1781, the fifth incloſure is made paſture; for which it is prepared, by ſowing paſture graſs-ſeeds with the barley of the year 1780. And in this manner may the rotation be carried on without end. Here the labour is equally diſtributed; and there is no hurry nor confuſion. But the chief property of this rotation is, that two culmiferous or white-corn crops, are never found together: by a due mixture of crops, the ſoil is preſerved

preſerved in good heart without any adventitious manure. At the ſame time, the land is always producing plentiful crops: neither hay nor paſture get time to degenerate. The whole dung is laid upon the fallow.

Every farm that takes a graſs crop into the rotation muſt be incloſed, which is peculiarly neceſſary in a clay ſoil, as nothing is more hurtful to clay than poaching.

Rotation in a free ſoil.

Incloſ.	1775.	1776.	1777.	1778.	1779.	1780.
1.	Turnip.	Barley.	Hay.	Oats.	Fallow.	Wheat.
2.	Barley.	Hay.	Oats.	Fallow.	Wheat.	Turnip.
3.	Hay.	Oats.	Fallow.	Wheat.	Turnip.	Barley.
4.	Oats.	Fallow.	Wheat.	Turnip.	Barley.	Hay.
5.	Fallow.	Wheat.	Turnip.	Barley.	Hay.	Oats.
6.	Wheat.	Turnip.	Barley.	Hay.	Oats.	Fallow.
7.	Paſture.	Paſture.	Paſture.	Paſture.	Paſture.	Paſture.

For the next rotation, the ſeventh incloſure is taken up for corn, beginning with an oat-crop, and proceeding in the order of the fourth incloſure; in place of which, the third incloſure is laid down for paſture, by ſowing paſture-graſſes with

with the laſt crop in that incloſure, being barley. This rotation has all the advantages of the former. Here the dung is employed on the turnip-crop.

We proceed to conſider what rotation is proper for carſe clay. The farm I propoſe conſiſts of ſeventy-three acres. Nine are to be incloſed for a kitchen-garden, affording plenty of red clover to be cut green for the farm-cattle. The remaining ſixty-four acres are divided into four incloſures, ſixteen acres each, to be cropped as in the following table.

Incloſ.	1775.	1776.	1777.	1778.
1.	Beans.	Barley.	Hay.	Oats.
2.	Barley.	Hay.	Oats.	Beans.
3.	Hay.	Oats.	Beans.	Barley.
4.	Oats.	Beans.	Barley.	Hay.

Here the dung ought to be applied to the barley.

Many other rotations may be contrived, keeping to the rules above laid down. For a clay ſoil, fallow, wheat, peaſe and beans, barley, cabbage, oats. Here dung muſt be given both to the wheat and cabbage. For free ſoil, drilled turnip, barley, red clover, wheat upon a ſingle furrow,

furrow, drilled potatoes, oats. Both the turnip and potatoes muſt have dung. Another for free ſoil: turnip drilled and dunged, barley, red clover, wheat on a ſingle furrow with dung, peaſe, barley, potatoes, oats. The following rotation has proved ſucceſsful in a ſoil proper for wheat. 1. Oats with red clover after fallow, without dung. 2. Hay. The clover-ſtubble dunged, and wheat ſown end of October with a ſingle furrow. 3. Wheat. 4. Peaſe. 5. Barley. Fallow again. Oats are taken the firſt crop to ſave the dung for the wheat. Oats always thrive on a fallow, though without dung; which is not the caſe of barley. But barley ſeldom fails after peaſe. In ſtrong clay ſoil, the following rotation anſwers. 1. Wheat after fallow and dung. 2. Beans ſown under furrow as early as poſſible. Above the beans, ſow peaſe end of March, half a boll per acre, and harrow them in. The two grains will ripen at the ſame time. 3. Oats or barley on a winter-furrow with graſs-ſeeds. 4. Hay for one year or two; the ſecond growth paſtured. Lay what dung can be ſpared on the hay-ſtubble, and ſow wheat with a ſingle furrow. 5. Wheat. 6. Beans or peaſe. 7. Oats. Fallow again.

CHAP.

CHAP. VIII.

Reaping Corn and Hay Crops, and Storing them for use.

Culmiferous plants are ripe when the ſtem is totally white: they are not fully ripe if any green ſtreaks remain. Some farmers are of opinion, that wheat ought to be cut before it is fully ripe. Their reaſons are, firſt, that ripe wheat is apt to ſhake; and, next, that the flour is not ſo good. With reſpect to the laſt, it is contrary to nature, that any ſeed can be better in an unripe ſtate, than when brought to perfection: nor will it be found ſo upon trial. With reſpect to the firſt, wheat, at the point of perfection, is not more apt to ſhake than for ſome days before: the huſk begins not to open till after the ſeed is fully ripe; and then the ſuffering the crop to ſtand becomes tickliſh: after the minute of ripening, it ſhould be cut down in an inſtant, if poſſible.

This leads to the perſons that are commonly engaged to cut down corn. In this country, the univerſal practice was, to provide a number of hands, in proportion to the extent of the crop, without regard to the time of ripening. By this method, the reapers were often idle for want of work; and what is much worſe, they had often more work than they could overtake, and ripe fields were laid open to ſhaking winds. The Lothians have long enjoyed weekly markets for reapers, where a farmer can provide himſelf with the number he wants; and this practice is creeping into neighbouring ſhires. Where there is no opportunity of ſuch markets, ought not neighbouring farmers to agree in borrowing and lending their reapers? The advantage is obvious; and yet I believe it is ſeldom practiſed.

One ſhould imagine, that a caveat againſt cutting corn when wet, is unneceſſary; yet from the impatience of farmers to prevent ſhaking, no caveat is more ſo. Why do they not conſider, that corn ſtanding, dries in half a day, when in a cloſe ſheaf, the weather muſt be favourable if it dry in a month? in moiſt weather it will never dry.

With reſpect to the manner of cutting, I muſt premiſe, that barley is of all the moſt difficult grain to be dried for keeping. Having no huſk, rain has eaſy acceſs; and it has a tendency to malten when wet. Where the ground is properly ſmoothed by rolling, I am clear for cutting it down with the ſcythe. This manner being more expeditious than the ſickle, removes it ſooner from the danger of wind; and gives a third more ſtraw, which is a capital article for dung, where a farm is at a diſtance from other manure. I except only corn that has lodged; for there the ſickle is more convenient than the ſcythe. As it ought

ought to be dry when cut, bind it up directly: if allowed to lie any time in the ſwath, it is apt to be diſcoloured.

Barley ſown with graſs-ſeeds, red clover eſpecially, requires a different management. Where the graſs is cut along with it, the difficulty is great of getting it ſo dry as to be ventured in a ſtack. The cunning way is, to cut the barley with a ſickle above the clover, ſo as that nothing but clean barley is bound up. Cut with a ſcythe the ſtubble and graſs: they make excellent winter-food. The ſame method is applicable to oats; with this only difference, that when the field is expoſed to the ſouth-weſt wind, it is leſs neceſſary to bind immediately after mowing. As wheat commonly grows higher than any other grain, it is difficult to manage it with the ſcythe; for which reaſon the ſickle is preferred in England. Peaſe and beans grow ſo irregularly, as to make the ſickle neceſſary.

The beſt way for drying peaſe, is to keep ſeparate the handfuls that are cut: though in this way they wet eaſily, they dry as ſoon. In the common way of heaping peaſe together for compoſing a ſheaf, they wet as eaſily, and dry not near ſo ſoon. With reſpect to beans, the top of the handful laſt cut, ought to be laid on the bottom of the former; which gives ready acceſs to wind. By this method, peaſe and beans are ready for the ſtack in half the ordinary time.

The

The ſize of the ſheaves ought to be regarded. A ſheaf commonly is made as large as can be contained in two lengths of the corn made into a rope. To ſave frequent tying, the binder preſſes it down with his knee, and binds it ſo hard as totally to exclude air. If there be any moiſture in the crop, which ſeldom fails, a proceſs of fermentation and putrefaction commences in the ſheaf; which is perfected in the ſtack, to the deſtruction both of corn and ſtraw. How barbarous and ſtupid is it, to make the ſize of a ſheaf depend on the height of the plants! By that rule, a wheat-ſheaf is commonly ſo weighty, as to be unmanageable by ordinary arms: it requires an effort to move it, that frequently burſts the knot, and occaſions loſs of grain, beſide the trouble of a ſecond tying. I have long practiſed the following method with ſucceſs. My ſheaves are never larger than to be contained in one length of the plant, cut cloſe to the ground: I admit no exception if the plants be above eighteen inches high. The binder's arm compreſſes the ſheaf ſufficiently, without need of his knee. The additional hands that this way of binding may require, are not to be regarded, compared with the advantage of drying ſoon. Corn thus managed may be ready for the ſtack in a week: it ſeldom in the ordinary way requires leſs than a fortnight, and frequently longer. Of a ſmall ſheaf compreſſed by the arm only, the air pervades every part;

nor is it ſo apt to be unlooſed as a large ſheaf, however firmly bound. The ordinary practice of directing the ſhocks to the ſouth-weſt for reſiſting the force of that wind, muſt be approved: but I cannot approve the placing on each ſide five large ſheaves, ſuch as require for a binding two lengths of the corn; which make ſo long a line as to be but imperfectly covered with the two head ſheaves. There ought to be no more but four ſheaves on a ſide. Five of my ſmall ſheaves, occupying ſtill leſs ſpace than four of the ordinary ſort, are covered ſufficiently by the two head ſheaves; and for that reaſon, I follow the ordinary practice of twelve ſheaves to a ſhock.

Every article is of importance that haſtens the operation, in a country like Scotland, ſubjected to unequal harveſt-weather. For carrying corn from the field to the ſtack-yard, a ſledge is a very aukward machine: many hands are required, and little progreſs made. Waggons and large carts are little leſs dilatory, as they muſt ſtand in the yard till unloaded ſheaf by ſheaf. My way is, to uſe long carts moveable upon the axle, ſo as at once to throw the whole load on the ground; which is forked up to the ſtack by a man appointed for that purpoſe. By this method, two carts will do the work of four or five.

It will not be eaſy to convince me, that building round ſtacks in the yard is not preferable to houſing the corn. Here it is ſhut up from the air;

air; and it muſt be exceedingly dry, if it contract not a muſtineſs, which is the firſt ſtep to putrefaction. Let me add another circumſtance, which would make a figure were it detached from that now mentioned. In the yard, a ſtack is preſerved from rats and mice by being ſet on a pedeſtal: no method has hitherto been invented, for preſerving corn in a houſe from ſuch deſtructive vermin *. The proper manner of building, is to make every ſheaf incline downward from its top to its bottom. Where the ſheaves are laid horizontally, the ſtack will take in rain both above and below. The beſt form of a ſtack is that of a cone placed on a cylinder; and the top of the cone ſhould be formed with three ſheaves drawn to a point. If the upper part of the cylinder be a little wider than the under, ſo much the better.

The

* Magazines for corn have been much extolled both in France and in England. But beſide the immoderate expence, they are very prejudicial to the commerce of corn. A farmer who has his ſtacks upon pedeſtals, waits patiently for a market; and thereby the price of corn is regulated by the demand. The proprietors of magazines, who are few in number compared with farmers, can by combination fix the price above or below the demand as it ſuits their intereſt; which is hurtful to buyers; and ſtill more to poor tenants, who, if it were not for magazines, would draw high prices in caſe of a ſcanty crop. This would be a great diſcouragement to agriculture, and make the farmer relax from his induſtry. Many would abandon the buſineſs altogether.

The delaying to cover a ſtack for two or three weeks, though common, is however wonderfully abſurd; for if much rain fall in the interim, it is beyond the power of wind to dry the ſtack. Vegetation, begun in the external parts, ſhuts out the air from the internal; and to prevent a total putrefaction, the ſtack muſt be thrown down, and expoſed to the air, every ſheaf. In order to have a ſtack covered the moment it is finiſhed, ſtraw and ropes ought to be ready; and the covering ought to be ſo thick as to be proof againſt rain.

Scotland is ſubject not only to floods of rain, but to high winds. Good covering guards againſt the former, and ropes artfully applied guards againſt the latter. I will anſwer for the following mode. Take a hay-rope well twiſted, and ſurround the ſtack with it, two feet or ſo below the top. Surround the ſtack with another ſuch rope immediately below the eaſing. Connect theſe two with ropes in an up and down poſition, diſtant from each other at the eaſing about five or ſix feet. Then ſurround the ſtack with other circular ropes parallel to the two firſt mentioned, giving them a twiſt round every one of thoſe that lie up and down, by which the whole will be connected together in a network. What remains is, to finiſh the two feet at the top of the ſtack. Let it be covered with bunches of ſtraw laid regularly up and down; the low part

part to be put under the circular rope firſt mentioned, which will keep it faſt, and the high part be bound by a ſmall rope artfully twiſted, commonly called *the crown of the ſtack*. This method is preferable to the common way of laying long ropes over the top of the ſtack, and tying them to the belting-rope; which flattens the top, and makes it take in rain. A ſtack covered in the way here deſcribed, will ſtand two years ſecure both againſt wind and rain; a notable advantage in this variable climate *. So much for corn. Now for hay.

The

* A granery for holding the whole product of a farm, muſt be a very expenſive building, and a ſevere tax upon huſbandry. I have heard it computed, that laying aſide towns and villages, the Engliſh barns have coſt more money than all their other houſes together. I can eaſily conceive an indolent practice ſupported long by cuſtom againſt the cleareſt light. But expenſive works are ſeldom attempted but from neceſſity; and I cannot eaſily conceive what at firſt produced that expenſive mode of preſerving corn and ſtraw together, when both can be preſerved in good condition by ſtacks in the yard; and at any rate, when it is much leſs expenſive to ſtore up the grain ſeparated from the ſtraw. There was a time not long paſt, when the moſt inventive heads in England and France were employed upon contriving granaries for corn, but without ſucceſs. I have no difficulty to pronounce, that a ſtack built as above directed and ſet upon a pedeſtal, is the beſt way for preſerving corn, and that for years, far beyond the moſt complete granary that ever was contrived, even laying aſide the expence of building and of management.

The great aim in making hay is, to preſerve as much of the ſap as poſſible. All agree in this; and yet differ widely in the means of making that aim effectual. To deſcribe all the different means, might be profitable to the bookſeller; but the reader would loſe patience, and gather no inſtruction. I ſhall therefore confine myſelf to what I think the beſt. A crop of ryegraſs and yellow clover ought to be ſpread as cut. Let it lie a day or two; and in the forenoon after the dew is evaporated, rake it into a number of parallel rows along the field, termed *wind-rows*, for the convenience of putting it up into ſmall cocks. After turning the rows once and again, make ſmall cocks weighing a ſtone or two. At the diſtance of two days or ſo, put two cocks into one, obſerving always to mix the tops and bottoms together, and to take a new place for each cock, that the leaſt damage poſſible may be done to the graſs. Proceed in putting two cocks into one, till ſufficiently dry for tramp-ricks of 100 ſtone each. The eaſieſt way of erecting tramp-ricks, is to found a rick in the middle of the row of cocks that are to compoſe it. The cocks may be carried to the rick by two perſons joining arms together. When all the cocks are thus carried to the rick within the diſtance of forty yards or ſo, the reſt of the cocks will be more expeditiouſly carried to the rick, by a rope wound about them and dragged by a horſe. Two ropes are

are ſufficient to ſecure the ricks from wind, the ſhort time they are to ſtand in the field. In the year 1775, ten thouſand ſtone were put into tramp-ricks the fourth day after cutting. In a country ſo wet as many parts of Scotland are, expedition is of mighty conſequence in the drying both of hay and corn.

With reſpect to hay intended for horned cattle, it is by the generality held an improvement, that it be heated a little in the ſtack. But I violently ſuſpect this doctrine to have been invented for excuſing indolent management. An ox, it is true, will eat ſuch hay; but I have always found that he prefers ſweet hay; and it cannot well be doubted, but that ſuch hay is the moſt ſalutary and the moſt nouriſhing.

The making hay conſiſting chiefly of red clover, requires more care. The ſeaſon for cutting is the laſt week of June, when it is in full bloom: earlier it may be cut, but never later. To cut it later, would indeed produce a weightier crop; but a late firſt cutting makes the ſecond alſo late, perhaps too late for drying. At the ſame time, the want of weight in an early firſt cutting, is amply compenſated by the weight of the ſecond.

The additional labour required to make hay of red clover, ariſes from the largeneſs of the ſtem, and the hazard of the leaves dropping off in moiſt weather. I have tried two methods. One

One is, to let it lie in the ſwath two days, and longer if the weather be unfavourable. The ſwaths muſt be turned over and over two or three times every day, but not unleſs the weather be dry. It will then be ready to be put into cocks, containing each about two ſtone. After two, three, or four days, according to the weather, let two cocks be put into one; and ſo on at proper intervals till ready for the tramp-rick. The other way, more expeditious, may be ventured on where ryegraſs is mixed with the clover. Stir it not the day it is cut. Turn it in the ſwath the forenoon of the next day; and in the afternoon put it up in ſmall cocks. The third day put two cocks into one, enlarging every day the cocks till they be ready for the tramp-rick. Sixteen pounds of red clover cut in the bloom, are reduced to four pounds when ſufficiently dry for keeping. I have tried, but without ſucceſs, to prepare it for keeping with a leſs diminution of weight. Ryegraſs cut in the bloom loſes of weight the ſame proportion; which was contrary to my expectation.

When the ſeaſon is too variable for making hay of the ſecond growth, mix ſtraw with that growth, which will be a ſubſtantial food for cattle during winter. This is commonly done by laying ſtrata of the ſtraw and clover alternately in the ſtack. But this method I cannot approve: if the ſtrata of clover do not heat, they turn mouldy

mouldy at leaſt, and unpalatable. The better way is to mix them carefully with the hand before they are put into the ſtack. The dry ſtraw imbibes moiſture from the clover, and prevents heating.

I muſt add in general with reſpect to hay of whatever kind, that if the weather be ſo wet as to prevent cocking in the ordinary time, there ought to be no intermiſſion in turning the ſwaths; which will help to evaporate the moiſture, if there be any motion in the air; and at the ſame time, prevent the ſwaths from ſinking into the ground among the uncut graſs, which never fails to blanch it.

The expence of making an acre of hay, compoſed of ryegraſs and yellow clover, is, in a tolerable ſeaſon, from four to ſix ſhillings; and from ſix to eight if compoſed of red clover. This however is an uncertain computation, the expence differing greatly in a good or bad ſeaſon.

I will not ſtop to give any rule for making hay of natural graſs, termed *meadow-hay*; becauſe in a well-conducted farm there ought to be no meadow-hay. This is made evident in the chapter *Rotation of Crops*.

In the yard, a ſtack of hay ought to be an oblong ſquare, if the quantity be greater than to be eaſily ſtowed in a round ſtack; becauſe a ſmaller ſurface is expoſed to the air, than in a number of round ſtacks. For the ſame reaſon, a ſtack of

 peaſe

peaſe ought to have the ſame form, the ſtraw being more valuable than that of oats, wheat, or barley. The moment a ſtack is finiſhed, it ought to be covered; becauſe the ſurface-hay is much damaged by withering in dry weather, and by moiſtening in wet weather. Let it have a pavilion-roof; for more of it can be covered with ſtraw in that ſhape, than when built perpendicular at the ends. Let it be roped as directed above for corn-ſtacks; with this difference only, that in an oblong ſquare the ropes muſt be thrown over the top, and tied to the belt-rope below. This belt-rope ought to be fixed with pins to the ſtack: the reaſon is, that the ropes thrown over the ſtack will bag by the ſinking of the ſtack, and may be drawn tight by lowering the belt-rope, and fixing it in its new poſition with the ſame pins.

The ſtems of hops, being long and tough, make excellent ropes; and it will be a ſaving article, to propagate a few hop plants for that very end.

A ſtack of ryegraſs-hay, a year old and of a moderate ſize, will weigh, each cubic yard, eleven Dutch ſtone. A ſtack of clover-hay in the ſame circumſtances weighs ſomewhat leſs.

I conclude this article with obſerving, that till lately the making hay was little underſtood in Scotland; nor to this day is it generally underſtood. The method was, to expoſe it as much as poſſible

poſſible to the ſun and wind, and never to give it reſt till it was ſo dry as to be grindable in a mill, and rendered a *caput mortuum*. The poor animals reduced to ſuch food, were truly to be pitied: is it wonderful that they were too feeble for work? But this made no impreſſion, as the feebleneſs of the cattle was but in proportion to the lazineſs of the men. They had no check of conſcience for being idle, becauſe they were educated in that way, and knew no better.

CHAP. IX.

FEEDING FARM-CATTLE.

HAVING diſcuſſed the management of corn and hay, what naturally follows is, to apply them to the maintenance of farm-cattle: for to conſider corn as the food of human beings, falls not properly under any branch of agriculture.

As in this chapter are contained many different matters upon which the profit of a farm greatly depends, I wiſh what I have to ſay may be clearly apprehended. In that view the chapter is divided into five ſections. Green food is the ſubject of the firſt; dry food, of the ſecond; feeding for the butcher, of the third: the fourth contains rules for the wintering of cattle that are not

not intended for immediate ſale; and the fifth, rules for buying and ſelling cattle and corn.

1. Green Food.

I begin this ſection with the ſummer-food of farm-horſes. The manner of feeding them during ſummer, is and has been various through Scotland: none of them good. Some time ago, horſes were fed in balks between ridges of corn; which required the attendance of men, and waſted much time. In many parts, horſes are reduced to thiſtles; the time of the men being conſumed in pulling, and of the horſes in eating. In ſome places, a part of the common paſture is reſerved for them, termed *hained graſs.* The man appointed to attend them falls aſleep, and ſuffers the horſes to treſpaſs on the corn. Dogs are employed to chaſe them from it: they run about; and their fatigue is little leſs than when at work. To prevent this, the horſes are ſometimes tethered on the hained graſs: the half is loſt, being trodden under foot; beſide that they often break looſe, and deſtroy the ſtanding corn. The leaſt exceptionable is a graſs-incloſure; and yet far from deſerving approbation. In the firſt place, where the graſs is ſo rank as to afford plenty of food to the horſes in the intervals of work, a fourth part at leaſt is trodden under foot; the horſes beſide are peſtered in hot weather

ther with flies, and cannot feed with eafe. In the next place, they have no time for refting; and much time is loft in laying hold of them for the yoke. Laftly, few inclofures in the hands of a tenant, are in fo good order as to keep in horfes when they fee corn; and if they once break out, it is in vain to think of imprifoning them after. The approved method to prevent every inconvenience, is to feed the horfes with cut grafs, under cover. In the interval between the work of the forenoon and afternoon, they can fill the belly in an hour, and have time to reft another hour: nor is a moment loft in yoking.

Several plants ferve this purpofe, fainfoin, lucern, red clover, white clover, ryegrafs. Red clover is the beft. Sainfoin and lucern early cut are excellent food; but when they turn ftrong and reedy, horfes are not fond of them. The difadvantage of white clover and ryegrafs, is, that, being fmall plants, they are not eafily collected in heaps for food. Red clover is extremely luxuriant: it is eafily collected: and it ought to be eafily collected; for to feed properly a horfe of a middle fize, requires ten ftone a-day. It flowers the firft week of June; but in rich foil it will rife to eighteen inches before flowering. So rapid is its growth, that in good foil it may be cut thrice in a year, and afford over and above fome pafture. It fhould be cut in the morning when moift with dew:

dew: it is leſs palatable when cut dry. The cutting ought to begin long before flowering; that all may be cut before it is too old, and that it may grow the faſter for a ſecond cutting. Theſe conſiderations are not ſufficiently attended to: people are loath to cut till the clover is fully grown; tho' early cutting will upon the whole afford more food, as well as more palatable food, to horſes eſpecially, which diſlike old clover. I diſpair not to ſee all the corn-farmers in Scotland, depending on red clover for the ſummer-food of their cattle; and then we ſhall no longer be ſtunned with loud complaints commonly thrown out as excuſes for idleneſs: "How can I improve, having no food "for my horſes but bare lea or thiſtles? they "cannot work on ſuch food; I cannot ſtir a "foot." A horſe works as he is fed: it is ſurpriſing what work he will perform upon cut clover, without loſing fleſh. Many a ſummer, for ſeven or eight weeks running, have my horſes been daily employed in bringing lime from a quarry fifteen Engliſh miles diſtant, fed on red clover only; and at the end of the ſeaſon, as plump and hearty as at the beginning. Let another article be conſidered. In every farm, a great proportion is left out for paſture, if that can be called paſture which affords little or no food. A ſingle acre of good red clover, will give more food than fifteen or twenty ſuch acres. How much

much better might theſe acres be employed in bearing profitable crops of corn ?

But a ſkilful farmer will not confine himſelf to red clover for ſummer-food. There are other graſſes that ſpring more early, and grow later than red clover. Ryegraſs, ribwort, and yellow clover, flower a month before red clover, are fully ready for cutting green the middle of May, and if cut at that time continue growing, and may be cut a fortnight after red clover is gone. To enlarge the period of green food for cattle, is a deſireable object in huſbandry. It affords plenty of food both early and late ; it ſaves paſture-fields in ſpring till the graſs cover the ground, which retains the dew, and ſhelters the graſs-roots from withering winds ; and it enables the farmer to leave his fields rough at the end of the ſeaſon to keep out the froſt during winter, inſtead of eating them bare, which is the ordinary practice. Now theſe ſalutary effects, all of them, may be procured by a very ſimple operation ; which is, to ſow part of the field with ryegraſs, ribwort, and yellow clover mixed. Theſe plants are ready for being cut the middle of May ; and if the ſeaſon prove favourable, they may be cut again as late as even the middle of November. Cut graſs is of all the cheapeſt food, and the moſt agreeable to horſes. Therefore to add a month of this food, is a valuable improvement, eſpecially during ſpring : for however nouriſhing dry food may be, be, yet a horſe put upon green food turns remarkably more agile and plump *.

I proceed to the ſummer-food of horned cattle. A beaſt that chews the cud, takes in at once a large

* It may be agreeable to bring under one view the graſſes mentioned above, with reſpect to the ſeaſon of their ſhooting and ripening.

Ryegraſs ſown in ſpring with corn, ſhoots the year following, from the 20th to the 30th of April, according to the ſoil and ſeaſon. It flowers from the 1ſt to the 10th of June. The flower continues about eight days, falls off, and the ſeed at that time begins to form. The ſeed is ripe between 1ſt and 10th July. When it is fully formed, the ſtalk begins to turn brown; and more and more ſo till the ſeed fall, which is about the end of July.

Ribwort ſhoots the laſt week of April. The head in a thriving plant is three inches long, full of ſeed, which is completely ripe about the 10th of July. Upon the head are found at the ſame time ſeed formed, flowers, and part that has not yet flowered.

Red clover ſhoots from the 1ſt to the 10th of June; and in eight days after begins to flower. It continues in flower twenty days; and about the end of July the ſeed is ripe. The progreſs of white clover is preciſely the ſame.

Yellow clover ſhoots the laſt week of April, and flowers till Auguſt. On a ſtalk is found at the ſame time, ſeed ripe, ſeed half ripe, flowers, and ſhoots juſt beginning. It accordingly reſembles peaſe, and grows a long time. Its continual growing keeps the ſtem full of ſap. The time of cutting plants for hay is in the middle of their flowering; but as yellow clover flowers much longer than the others mentioned, it affords a greater latitude for cutting without injuring the hay.

large quantity of green food, especially of red clover, which is extremely palatable when young. So large a quantity is apt to ferment with the heat of the stomach, so as sometimes to make the creature burst. This is considered as a formidable objection to the feeding horned cattle on red clover. But it is easily obviated, by feeding them in the house: servants will not readily give more than sufficient, when cutting and carrying is a work of labour. And red clover should always be cut for food; for where cattle have liberty to pasture, more is trampled down than is eaten. At any rate, bursting may be prevented even when cattle are allowed to pasture. Indulge them but half an hour or so, for two or three days when the clover is dry; after which there is no hazard. If yellow clover and ribwort be sown with red clover, there is little or no hazard of fermenting to such a degree as to be hurtful. White clover is no remedy: it ferments in the stomach as much as red clover.

Red clover cut green, is preferable to all other food for milk-cows. Being soon filled, they have much time to rest, which increases the quantity of milk. The milk at the same time is richer and higher coloured, than from any other food.

Red clover is good food for sheep; but the cutting it for them would be too expensive. White clover at the same time is their favourite, which is never wanting in good soil, growing naturally.

One ſignal advantage of feeding horſes and horned cattle in the houſe during ſummer, is their being protected from heat and inſects. And it is a ſtill more ſignal advantage, that the dung turns to much better account, than when ſcattered during ſummer in a paſture-field. Horſe-dung in a paſture-field is totally loſt: it dries, and withers away, not to mention that its heat burns the graſs it falls on. Dung is an article of great importance, eſpecially in a farm diſtant from other manure. And a dunghill, procured by feeding on cut graſs, may be conſiderably increaſed by adding to the heap every weed that grows in the farm; which at any rate ought to be cut, to prevent ſeeding.

The carrying cut graſs from the field to the ſtable is a laborious work; and the only circumſtance that weighs againſt cut graſs in competition with paſture. A horſe of a middle ſize will eat ten Dutch ſtone daily; ſome go the length of ſeventeen: an ox or a cow will eat eight ſtone. Suppoſing in a farm ten horſes, ten oxen, and ſix cows: they will conſume 228 ſtone a-day. If the clover be at any diſtance, that quantity requires a cart going continually from morning till evening. Computing a cart at three ſhillings *per* day, the expence for the ſix ſummer-months is no leſs than L. 25, 4s. Even this high expence, is far from counterbalancing the advantage of feeding cattle in the houſe. The expence, however, is ſo conſiderable,

derable, as to make it of importance to leſſen it. In that view, I recommend the following plan, the purpoſe of which is, to carry the cattle to their food, inſtead of carrying food to the cattle. Erect a moveable ſhed in the field, all of wood, the back conſiderably higher than the front, in order to have a ſlopping roof againſt rain. Sixteen feet in wideneſs is ſufficient for a beaſt of any ſize; the length correſponding to the number of cattle that are to be fed. On the back at the heads of the cattle, a deal is hung with hinges, to be lifted up for throwing food to them. Upon the ground along the length of the houſe, three beams are laid, croſſed with ſpars an inch diſtant from each other. The channel or gutter behind the cattle, is lined with a deal in the bottom, and one on each ſide, to convey the urine from the cattle to a pit filled with rich earth; which I hold to be preferable even to dung itſelf. The three beams covered with ſpars make a vacuity below, which receives the urine at the firſt inſtance, and preſerves the cattle dry. This is an important article in feeding cattle, which every animal at liberty is fond to procure to itſelf. There is no neceſſity for racks in this ſhade; on the contrary, horſes eat more conveniently in the natural way, by bending down the neck to food. The only thing neceſſary is a board between their fore feet and the clover, to ſave it from being trampled under foot. It is proper that in a corner a bed

be

be erected for a ſervant, to attend the cattle during night. The deals of this ſhade muſt be held together with wooden nails, ſo as eaſily to be taken down, and ſet up again where it may be wanted. It ſhould be placed at the loweſt part of the field; becauſe the clover is more weighty, than the dung it produces. It is poſſible to ſet a little ſhed on wheels, to be carried from place to place, without being taken down; but that will never anſwer for a ſhed of thirty or forty yards long. Such a ſhed is proper in every farm, where red clover is annually raiſed in the courſe of cropping; and the expence will be the leſs grudged, conſidering that it alſo anſwers for conſuming turnip and cabbage in winter. Let the expence be computed of carrying theſe to an immoveable ſhed; and in a farm of any extent, it will be found that the expence thus ſaved, even in a ſingle year, will equal the coſt of the propoſed ſhed. In a ſmall farm, where red clover enters not into the rotation of crops, and yet is neceſſary for ſummer-feeding; the moſt convenient way is, to incloſe ſix or ſeven acres as near as may be to the farm-offices, upon a part of which there ſhould always be a crop of red clover in rotation. In that caſe the carriage is a trifle. I have only to add, that room ſhould not be ſpared; for horſes are hurt as much as horned cattle by being crowded.

The

The proper ſeaſons for diſpoſing of cattle fatted on graſs, are the June markets, and thoſe of December and January. With a view to the firſt, early graſs ought to be provided, and late graſs with a view to the others. In an open winter, there is no difficulty to preſerve graſs-cattle fat through December and January: in a hard winter, the addition of a little hay will do.

Next in order is the feeding cattle in a paſture-field. White clover is for paſture the beſt graſs known in Britain, being extremely palatable to cattle of every kind. It is a native of Britain; and like a ſtrawberry it throws out flagellæ or runners from the ſtem, which take root and become new plants. But as this is a work of time, it is more profitable to ſtock the field with it at once. Good ſeed is weighty, and full without dimples: it is red when hurt in drying. A field intended for paſture, requires a mixture of graſſes: every ſpecies of animals has its favourite graſs; and when animals of different ſpecies feed together, not a ſingle ſtump is left. Different graſſes alſo, having different times of flowering, keep the chain of food more complete during the ſeaſon. Ryegraſs for example anſwers ſpring-food better than any other plant, and continues longer after autumn. But white clover during ſummer holds out better than ryegraſs. Different graſſes at the ſame time excite

cite the appetite: an ox will leave turnip for cabbage; and after feeding plentifully on both, will take kindly to hay or ſtraw. The proper quantity of ſeed to an acre intended for immediate paſture, is ten pounds white clover, five pounds yellow clover, as much ribbed graſs, and two wheat-firlots of ryegraſs.

An incloſure proper for paſture, ought to have the following properties. It ought to be well aired. Second, well watered. Third, well ſheltered. Fourth, the larger the better. And, laſtly, the graſs ought to be ſo rank as to afford a full bite. With reſpect to the firſt, a field well aired makes cattle feed kindly: in a hot day, they go to the higheſt part for freſh air: if they have neither freſh air nor water to reſort to, they fret, and loſe fleſh. The want however may be ſupplied artificially, by clumps of evergreens ſcattered through the field, to ſhelter them from the ſun. With reſpect to the ſecond, plenty of water for drinking is not alone ſufficient: there ought to be plenty for bathing, in a hot day: cattle are never more at eaſe in ſuch a day, than when they are plunged in water. With reſpect to the third, it is not ſufficient that cattle be ſheltered againſt heat: ſhelter againſt cold is ſtill more neceſſary. By proper management, the chain of graſs may be carried on in tolerable weather till the end of the year; but unleſs the cattle be protected from cold blaſts, graſs will do them little good. The

clumps

clumps mentioned planted in the form of a crofs, will afford fhelter from whatever quarter the ftorm comes. With regard to the fourth, the field ought to be fo large as to give cattle their natural range. Every fpecies of animals that feed on grafs, have a natural range in feeding; and to confine them within narrower bounds, is to them a fort of imprifonment. Sheep have a wide range; and ought to have, becaufe they delight in fhort grafs: give them eighty or ninety acres, and any fence will keep them in: confine them to a field of feven or eight acres, and it muft be a very ftrong fence that keeps them in. A range of fifty or fixty acres is fufficient for horfes; and a ftill narrower range for horned cattle. In oppofition to the field defcribed, advert to cattle cooped up in a fmall inclofure of eight or nine acres, furrounded with high hedges. In fummer they are ftifled for want of air, are peftered with infects, and lofe fat inftead of gaining. To examine the progrefs of fattening, I weighed twenty ftots the firft day of May and the firft day of the five fucceeding months. Their quickeft advance was in May and September, being at an average two pounds daily each. I could not attribute this to any other caufe, but to lefs heat and fewer infects than in the three intermediate months. With refpect to the property laft in order, the benefit of a full bite is too obvious to need explanation.

A

A paſture-field bare of graſs in ſpring, having no protection againſt withering winds, turns hard and unfit for vegetation, eſpecially after wet weather. But where a field ſprings early, and is covered with graſs before drought ſets in, it continues moiſt and tender, by retaining dew and keeping out drought. To encourage early graſs in ſpring, the field ought to be left rough in winter; which keeps the ground warm, and protects the roots from froſt.

Ragwort is a troubleſome gueſt, as it never fails to infeſt rich paſture-fields: it is not only a robber, but overſhadows the graſs, and renders it unwholeſome. I am at a loſs however whether to call it a weed or an uſeful plant. As it bears no ſeed till the third year of its growth, it cannot propagate in land under tillage: in paſture-land, it dies indeed after dropping its ſeed, but new plants ſpring from that ſeed, and have a ſucceſſion without end. Many things in appearance noxious, have been found uſeful; of which this plant is an inſtance. The ſame means will prevent its noxious effects, and make it profitable. Ragwort in flower was never ſeen in a field paſtured with ſheep. Why? becauſe that animal is exceſſively fond of it. Therefore, in every paſture-field, for ſome years after it is laid down, there ought to be a proportion of ſheep. They prefer ragwort before any other vegetable; and experience pronounces, that every food is wholeſome which an animal

animal is fond of. Lincolnſhire ſheep do beſt; becauſe a fence ſufficient for horned cattle, is more than ſufficient for them. Sheep are ſingular with reſpect alſo to other food. They are fond of the tender ſhoots of broom and whins; and no leſs ſo of the fruit of the horſe-cheſnut.

The ruſh may be compared to ragwort: it is a troubleſome weed; and yet may be made in ſome degree profitable. Whether it ſhould be claſſed among the evergreens, appears doubtful from the following account of it. It ſprings ſix or ſeven inches high in April, grows on till Auguſt, when the ſeed appears at the ſide of the ſtem, five or ſix inches below the top. While the ſeed is drawing toward maturity, the part above withers gradually; and when the ſeed ripens and falls, the part below withers down till within a foot or ſo of the ground. That part continues green all winter, but dies away before next ſummer. In place of the ſtems that thus die, freſh ſtems ariſe, which, as obſerved above, make a figure in April. Whether theſe ſtems ariſe from ſeed, or from the bulky root, or from both, I cannot at preſent determine.

As the ruſh is an aquatic, and grows ſo vigorouſly as to deſtroy all other plants, it ought to be rooted out if poſſible; not only becauſe it is a bad paſture-graſs, but becauſe the infallible way of rooting it out is, to lay the ground dry; which makes a double improvement. But if this cannot be got done at a moderate expence, the reſource is, to make all the profit of it that is poſſible. Ruſhes cut in June while young and tender, and dried into hay, make tolerable winter-food: a freſh growth enſues which is proof againſt froſt, and is not unſavoury to cattle that run out all winter.

In the month of April, while graſs is ſtill ſo ſhort as not to afford a bite to horned cattle, I have ſeen fourteen cows living on ruſhes. At that ſeaſon, they are to horned cattle almoſt as palatable as red clover. They are rejected when old; and ſo is even red clover.

A ſprat is not an evergreen, for it dies away in winter. It is an aquatic, like a ruſh, but thrives with a leſs degree of moiſture. It may be of ſome uſe for hay when cut in June; and the after-growth is not unpalatable when eat young. But after feeding no beaſt will touch it.

2. Dry Food.

I proceed to dry food. As hay is of uſe to cattle of every kind, it is a capital object in huſbandry; and not the leſs capital, that, in my thought, it may be carried to greater perfection than is commonly done. To give ſatisfaction, we muſt enter into an examination of the different graſſes that are uſed for making hay. Red clover is a ſucculent plant of the leguminous tribe, that flowers

flowers about the beginning of June, and rifes to the height of between three and four feet. Rye-grafs is a culmiferous plant, dry and folid like other plants of that tribe; rifing commonly in good foil to three feet and above. It fhoots the end of April or beginning of May; and above all other plants haftens to perfect its feed. Yellow clover and ribwort, both of them leguminous plants, are more tender and fucculent than rye-grafs; but lefs fo than red clover: they rife commonly to three feet. They fhoot at the fame time with ryegrafs, but perfect their feed not altogether fo foon. Ryegrafs, becaufe of its folidity, is of all the beft food for horfes; and next to it, yellow clover and ribwort. Horned cattle delight in leguminous plants, red clover efpecially: it is probable that tender graffes are fitteft for animals that chew the cud. White clover blooms and perfects its feed at the fame time with red clover: it makes excellent dry food for fheep; but as it feldom rifes above eighteen inches, it is lefs proper for hay than any of the others mentioned; and upon that account, is often left out in the mixture of graffes for hay.

With refpect to the time of endurance, red clover cannot be depended on for hay longer than two years: the third year, it makes at beft a fcanty crop, and frequently vanifhes altogether. Ryegrafs, ribwort, and yellow clover, ftand good three years. From that time, the product leffens gradually;

gradually; and after the ſeventh or eigth year they afford little paſture.

The general practice of Britain for hay, has been to mix ryegraſs and red clover with a ſmall proportion of white clover; which I believe continues to be the practice in many places. I cannot help condemning it, as the child of ignorance or inattention. Two weighty objections occur. The firſt is, that when the red clover is fit to be cut for hay, which is about the beginning of July, the ryegraſs-ſeed is ripe. And what follows? The ryegraſs that year grows no more than barley or wheat cut when the ſeed is ripe; whereas if ryegraſs be cut before the ſeed is formed, it grows all ſummer, and even all winter, till it makes way for new ſhoots in ſpring. It may poſſibly be thought, that the ſeed, like that of other culmiferous plants, is more than ſufficient to make up the ſcantineſs of ſubſequent cuttings; but in drying for hay, moſt of it is loſt, ſome in the field, ſome in the ſtack, and ſome in the hay-loft; little being left but the dry ſtraw, which cattle do not willingly eat. The other objection is of ſtill greater weight. After a crop or two of hay, the field is ſurrendered to paſture. The red clover wears out the ſecond year, leaving nothing but the ryegraſs, which continues but another year in perfection: and the ſmall proportion of white clover has not yet had time to ſpread over the field. Thus, after the ſecond year, the paſture

ſture is but indifferent; and turns worſe and worſe till the field have time to ſtock itſelf with natural graſſes. Nor is that all: while the field is but half-ſtocked with ſown graſſes, weeds take poſſeſſion of the vacuities, and accelerate the dwarfing of the ryegraſs and white clover. And if the red clover happen to be a ſcanty crop the ſecond year, it encourages weeds no leſs than a ſcanty crop of peaſe.

I venture to ſuggeſt a better plan for hay than any I have read of. I pronounce the beſt hay for horſes to be a mixture of ryegraſs and yellow clover, which flower together and are fit to be cut for hay the beginning of June, when ſtill in flower and the ſeed not formed: at that time, being full of ſap, they are in perfection for hay. When cut ſo early, they grow again vigorouſly, and afford a ſecond crop end of September; or excellent paſture, if the ſeaſon be unfavourable for hay. When the firſt cutting is later than beginning of June, the ſtem turns dry and woody; and much of it is rejected, if a horſe be not extremely hungry. But I recommend this to gentlemen only and farmers, for feeding their own horſes. In the view to ſell hay, their profit I acknowledge will be greater in cutting when the ſeed is ripe, becauſe the ſeed is a profitable article. The hay, it is true, ſeparate from the ſeed, is no better than ſtraw: but that is no objection to an innkeeper; for the leſs a travelling horſe

horſe conſumes, the profit is the greater. My reaſon for chuſing theſe two plants for hay to horſes is, that, of all leguminous plants, yellow clover approaches the neareſt in ſolidity to rye-graſs, and that they dry well together. Ribwort is left out, as being leſs fit for horſes; and with plenty of ſeed the two plants choſen will make as weighty a crop as the ground can bear. Three wheat-firlots of ryegraſs, and ten pounds of yellow clover, make a ſufficient doze for an acre. And when cut early, as it ought to be, the crop muſt be very weighty if the cutting coſt more than eighteen pence *per* acre.

For hay to horned cattle, the mixture ought to be red clover, yellow clover, and ribwort. Red clover, it is true, falls properly to be cut a little later than the other two; but as it is choice food for horned cattle, it is better to cut it along with the other plants, than to loſe their foggage by cutting them too late. Ten pounds red clover, ſix pounds ribwort, and four pounds yellow clover, are ſufficient for an acre.

A farm-horſe, during winter, requires at leaſt two lippies of oats daily: in ſpring, which affords more working-hours, they are encreaſed to three. The reſt of his food is hay or ſtraw: bean-ſtraw is the beſt, peaſe-ſtraw next, and after it good oat-ſtraw. Upon that food, a pair of good horſes will labour ten hours a-day, and plow an acre of cultivated land; or draw on a ſmooth road

a

a hundred ſtone in a cart, ſixteen Engliſh miles. Such work they can perform daily without loſing fleſh.

When too much hay or corn is given to a horſe at once, he eats greedily, and is apt to ſurfeit on it. On this account, as well as to ſave food, better to give it in ſmall parcels at intervals, in which caſe every particle will be eat up clean. Were it not for the expence, the beſt way of feeding a horſe with hay would be to give it out of the hand. As for corn, a horſe when hungry, is apt to ſwallow it without chewing; in which caſe it paſſes entire and gives little nouriſhment. Therefore with oats mix chaff or cut ſtraw, which require chewing. If that labour be grudged, there is an inſtrument contrived for bruiſing oats, which has a promiſing appearance. I will not, however, venture to recommend it, becauſe I have no experience of it.

A dairy may be turned to great account. A good cow, during the ſix ſummer-months, will give at a medium twelve Scotch pints of milk daily; the butter of which, with the ſkimmed milk, may amount to eighteen pence *per* day, and thirteen pounds ten ſhillings in the ſix months. The graſing of ſuch a cow for that time, will not coſt above forty ſhillings, which makes eleven pounds ten ſhillings of profit.

3. FEEDING

3. Feeding for the Butcher.

Horned cattle continue to grow till they are full ſix years old; during which time they do not readily take on fat, nor carry much tallow. The proper time to enter them upon fattening food, is at the age of ſeven. There is an additional reaſon, that three years work can be got from them, which in all events will do more than balance the expence of their food.

As the demand for butcher-meat in Scotland increaſes rapidly, the feeding cattle for the ſhambles has become an intereſting article to the farmer. Thirty years ago, he had no temptation to keep fat on his cattle during winter, becauſe ſalt meat was our only food during winter and ſpring. We have now freſh meat in plenty, all the year round; only a little dearer in ſpring; and of that circumſtance a provident farmer will avail himſelf. The difference of a month, will ſometimes add a halfpenny to the price of a pound of beef; which, upon an ox of ſixty ſtone, makes forty ſhillings for that month's feeding.

A field ſown with the graſſes above mentioned, may be depended on for fattening from the middle of April, in a warm ſoil, till the middle of November: nor will they fall off till the middle of December, if froſt and ſnow keep away. Hay, though greatly preferable to ſtraw

of

of any kind, yet will not fatten horned cattle for the market without green food. It indeed keeps the fat upon them; but adds little or none. It is however profitable to feed a fat ox with hay during winter; becauſe the additional price got in the ſpring, will defray the expence of the hay. Some years ago, I fed twelve oxen with hay through winter; and though they were not a pound heavier in April, I ſold them at that time for double the ſum I paid for them. An ox of a middling ſize, eats thirty pounds of hay, Dutch weight, in twenty-four hours, and drinks forty-eight Scotch pints of water. An ox of a large ſize will eat forty pounds, and will drink in proportion.

Of all the beaſts I know, a ſpayed quey is the moſt profitable both for labour and for the butcher. She is little inferior to an ox in ſtrength; but much more agile, and conſequently better fitted for travel. She will work even till eight years old, and ſtill be fit for feeding fat. People of delicate taſte prefer her beef.

It is a great loſs to have a cow to maintain that has miſſed calf. The only profit is to feed her for the butcher. She feeds the faſter after having received the bull. But care muſt be taken not to admit the bull, but ſo as that ſhe may be ſold fat three months before the time of calving; for during theſe three months ſhe commonly loſes fat.

Conſiderable profit may be made by feeding ſmall cattle bred in the Highlands, purchaſed when four years old at a ſmall price, becauſe they are unfit for work. This branch of commerce is accordingly well underſtood.

A ſow is a profitable animal: it feeds greedily on cut clover; which, with the offals of the kitchen, dairy, and barn, prepare it finely for being fatted in a ſhort time with more nouriſhing food.

The feeding of calves for veal is alſo profitable. The calf even of a middle-ſized cow, after ſucking her for ſix weeks, will ſell at thirty ſhillings. The ſame cow will in ſix months feed four calves for the butcher, ſometimes more; and after all, will give milk two months longer. For making fine veal, the calf ought to be blooded frequently, kept from light, laid clean and dry, and get chalk to lick.

For ſome years paſt, a ſhed erected upon pillars, with intervals of eight or ten feet, has been uſed for ſtall-feeding. Such a ſhed I pronounce to be too cold for our climate. A houſe ſo conſtructed as to avoid the extremities of heat and cold, anſwers much better. Upon a feed of turnip in hard froſt, cattle may be obſerved to contract their feet together, and to tremble as in an ague. No animal can feed well in diſtreſs. On the other hand, cattle in a cloſe houſe, and in a hot day, may be obſerved panting for want of breath. Therefore let the feeding-houſe have

many

many windows or air-holes, to be ſhut or opened as occaſion requires.

As to the time of houſing cattle for feeding, it is evident, that as graſs is more eaſily raiſed than cabbage or turnip, the longer ſtall-feeding is delayed without loſing fat, the better for the farmer.

With reſpect to the food proper for ſtall-feeding, turnip, cabbage, colewort, potatoes, carrot, are all proper, and may be raiſed in every farm. For feeding in perfection, all of them ought to be provided; for which there is more than one reaſon. In the firſt place, variety excites the appetite: next, ſome of theſe vegetables endure the winter better than others: turnip, for example, does not anſwer for ſpring-food, ſo well as cabbage and colewort. Therefore, as far as is conſiſtent with variety, I would be ſparing of the vegetables that are the leaſt hurt by winter. Potatoes anſwer beſt for the concluding food. If ſufficient ſtore be provided and well preſerved, which is an eaſy matter, all complaints of wanting green food in ſpring muſt be at an end. Every animal is fond of potatoes, not even excepting a horſe. They are a choice food for milk-cows, and produce plenty of milk; which has no rank taſte more than where fed on hay or graſs. And yet after all, how many indolent farmers ſtill remain, who for want of ſpring-food are forced to turn their cattle out to graſs, before it is ready for paſture;

paſture; which not only ſtarves the cattle, but lays the graſs-roots open to be parched by ſun and wind. One precaution, however, is neceſſary with reſpect to graſs in proper condition for paſture: ſurrender it not to the cattle at once, which would give a looſeneſs, but for a few days let them fed an hour or two only.

Preparatory to the feeding cattle in the houſe with cabbage and turnip, they ought to be made acquainted with that food in the field, by getting now and then ſmall feeds of it; beginning with cabbage, which they ſoon take to as being ſoft, and then proceeding to turnip. Without being thus prepared, ſome beaſts have been known to faſt obſtinately for days, before they would touch them. In the houſe, it is common to fill the ſtalls at once for ſaving trouble. This is wrong: cattle are ſo fond of cabbage and turnip after acquaintance, as to be apt to ſurfeit upon them. A hungry horſe, by eating too much corn at a time, takes a loathing of that food; and his loathing continues for days. Begin with giving a bullock one turnip after another, and he will not readily ſurfeit. One bullock of ninety ſtone Dutch thus fed, ate up thirteen Dutch ſtone before he ſtopped; and in twenty-four hours he devoured thirty ſtone. This was the third part of his own weight; and every bullock in health will do the ſame, when thus fed. Therefore, after cleaning the turnip, let the feeder begin in the morning with

with throwing a turnip or two to every bullock: let him repeat the dofe till they turn fhy: let him then give them cabbage in the fame manner; and conclude with handfuls of hay. When they are thus fatisfied, prepare them a foft bed as an invitation to lie down. Leave them to chew the cud and fleep, till they rife of their own accord. Repeat the dofe as before. As they grow fat, they will reject the coarfer parts of the food; which may be given to winterers, or to fwine. If the turnip be large, there is no danger of giving them whole. But they are apt to choke upon fmall turnip; which therefore muft be cut into fmall bits that can be eafily fwallowed. A machine to bruife turnip for eating would be an ufeful invention.

Thefe inftructions are intended to make cattle eat the greateft quantity poffible; it being certain, that the more a bullock eats with an appetite, the fooner he grows fat. It is bad œconomy to fpare food in this cafe: a certain quantity daily is requifite to preferve a beaft from falling away; and an addition is neceffary to put fat on him. Therefore the fooner he is fatted, the greater proportion of what is neceffary for bare maintenance, is faved.

To keep cattle clean and well littered, is to them half food; and this may be eafily done in a fhed conftructed as defcribed above. Let them be combed every day, wafhed with water every week,

week, and bled every month; for theſe all contribute to fattening.

Moſt of the cattle in this country ſold to the butcher, are fed upon graſs, and ſold in November at two pence *per* pound, or two pence halfpenny. But as the profit here is very ſmall, a farmer who is ſtudious of his intereſt, will have a proviſion of cabbage and turnip for winter, in order to ſell his cattle at double price in ſpring.

It has become a pretty general practice, to feed cattle in the field with turnip, cabbage, and other annual greens. There is not another way of fattening cattle ſo cheap; nor another way by which ſo much dung can be raiſed in a farm. Theſe are important objects. But what I chiefly inſiſt on is, that there is not within the invention of man, a more effectual method for improving a gravelly or ſandy ſoil. Nothing indeed is more hurtful to clay or rank loam than poaching; but the poaching a light ſoil takes away the pores, makes the earth more compact, and more retentive of moiſture. A crop of turnip or cabbage on ſtiff land, ought not to be fed in the field; but in a ſhed, ſuch as above deſcribed, or in an adjacent dry field.

To be ſucceſsful in this article, let cattle be choſen that have been accuſtomed to run out in winter. Cattle that have been always houſed in winter, are too tender for field-feeding: they will never turn fat.

There

There requires more management in feeding cattle with cabbage and turnip in the field, than is commonly practiſed. What muſt one think of the ſlovenly method of turning cattle into a turnip-field, to feed at random? I hope this is rare; but another method is not rare, which is, to incloſe with flakes a portion of the field, and within that ſmall ſpace to confine the cattle. Is it not obvious, that by dunging, piſſing, trampling, a great part of the turnip muſt be loſt, or become nauſeous? To prevent that loſs in a light porous ſoil that turns more ſolid by poaching, I have always practiſed the following method. Suppoſing the incloſure to be an oblong ſquare, which is the moſt convenient for flakes, begin at one of the ſhort ſides, and from the fence throw the turnip towards the middle of the field, clearing as much ground as can be done at one throw, which may be thirteen or fourteen feet. Separate this vacant ſpace from the turnip by flakes. Let the flakes incline inward to the field, which will prevent the cattle from rubbing them down. Introduce the cattle into this void ſpace, and begin with throwing over to them, from time to time, the turnip that were taken up, ſo ſparingly that they may eat without trampling them under foot. After theſe are clean ate up, clear another ſtrip of the ſame breadth with the former, by throwing over to the cattle the turnip that grow there. Remove the flakes to the ſide of the growing turnip,

turnip, and go on till the field be ate up. In this manner, the whole field will be knead and poached, ſo as totally to alter the texture of the ſoil. But becauſe to give the cattle no other bed, would greatly retard the progreſs of fattening; an adjacent graſs-field is neceſſary, in which they ſhould be put every night for a dry bed. In this graſs-field place hecks, for feeding the cattle with hay or ſtraw; as nothing contributes more to expeditious fattening, than alternate green and dry food.

Froſt is the only enemy to this manner of feeding. In froſt, the digging up the turnip is expenſive; and they give the cattle a looſeneſs, which makes them lean inſtead of fat. Hard froſt deſtroys the turnip altogether. I have tried ſeveral preventives. At the corner of a field, I built in a ſtack all the turnip that grew on four acres. They remained entire ſix weeks; but after that time, they began to ſprout, eſpecially about the centre, which exhauſted the turnip, and rendered them leſs palatable. I tried another method; which was, to fill a large houſe with turnip, on the firſt appearance of froſt. The warmth of the houſe made them ſprout ſooner than in the former way. A third experiment was to dig a pit in the field ſix or ſeven feet deep, where I ſtowed the turnip and covered them with three feet of earth, reckoning that this would prevent vegetation. But they ſprouted ed as ſoon as in the houſe. A fourth experiment was to build in the field what would fill two large carts, covering the heap with ferns or ſtraw. In this ſituation, they were longer of ſprouting; but they were more expoſed to froſt. Upon the whole, the beſt method, as appears to me, is to ſtore up in a houſe as many turnip as will ſerve the cattle a fortnight. As after that time they begin to wither, give them to the cattle though the weather be freſh; and then fill the houſe again. This is a reſource againſt a fortnight's froſt; and as a farther reſource, a few heaps may be collected in the field, ready at hand if froſt ſet in. A reſource ſtill more commodious of late invention is mentioned below page 208. The beſt preventive of all againſt a long froſt are potatoes, which may be eaſily preſerved in a houſe during the longeſt froſt. Cattle are fond of potatoes, and they make a good change of food to ſupply the want of turnip or cabbage during froſt. The ſame method may be taken with cabbage. Carrots are preſerved by taking them up, and burying them in ſand. Potatoes being more hurt by froſt than any other root, they ought to be ſtored up in a houſe, and covered with dry ſtraw preſſed down with litter. But here a precaution is neceſſary. If allowed to lie too long in their winter-quarters, they will ſprout, and be uſeleſs. Therefore, as ſoon as the froſt is over in ſpring, they muſt be expoſed to the ſun and air till perfectly dry; and then laid up where there is a free circulation of air.

Cabbage and turnip give a ſtrong taſte to butcher-meat, when not intermixed with dry food. But the remedy is eaſy: give nothing but dry food for a few days before they are delivered to the butcher. The intermixing dry food with turnip and cabbage in feeding milk-cows, ought never to be omitted; becauſe theſe plants without that intermixture give a nauſeous taſte to the milk, cream, and butter.

The feeding cattle with turnip and cabbage in the field, is an intereſting article. There are few farms but what are fit for raiſing green food of one kind or other; and by far the greater part are fit for every kind. Light ſoil is fit for turnip, both light and loamy ſoil for potatoes, deep ſoil for carrots, and ſtrong ſoil for cabbage. For a dozen of years back, the ordinary profit of ſuch feeding from the end of November to the end of March, has been to double the price of the cattle; which affords a very handſome rent for the land. Nor do I reckon this the capital part: the gain is ſtill greater by the improvement of the ſoil. No other method is ſo effectual to convert light porous ſoil into what is firm and heavy, as the poaching of cattle during winter. To compare it with ordinary manure, would not do it juſtice. It alters and improves the ſoil like clay-marle; and theſe two are the only means I am am acquainted with, to produce a new ſoil equal to one originally rich. There are few ſoils indeed but what may be hurt by cropping; but a ſoil ſo improved, can never be hurt ſo much by cropping, as to be reduced to its original poor ſtate. After ſuch a winter-feeding, it would ſurpriſe one, to find ground formerly ſo looſe as not to hold together, become now ſo ſolid as to be raiſed by the plough in ſolid maſſes. This way of feeding cattle, is at the ſame time a good dunging to the field, equal at leaſt to an ordinary dunging from a dunghill. The cattle evacuate abundantly, and what drops from them mixes intimately with the ſoil: not a particle is loſt.

The feeding ſheep on turnip and cabbage, ought to be delayed as long as poſſible, for the reaſon given above with reſpect to horned cattle. But as ſheep are better protected againſt cold, there is the leſs neceſſity of providing a ſhade for them. To ſave the turnip, flakes ought to be uſed; and a dry bed is ſtill more neceſſary to them than to horned cattle. Their dry food ought to be white-clover hay, or unthreſhed fitches; both of which they delight in. White-clover hay, however, without corn or green food, will not fatten ſheep for the butcher. In froſt and ſnow, ſheep ſhould always be ſupplied with hay or other dry food; which prepares them for green food the end of March; at which time a few

few weeks of ryegraſs will make them ready for the market, at the deareſt time of the year.

Where ſheep are fed in the houſe, which is ſeldom done except for the uſe of a family, they ought to be ſeparated by rails from each other; otherwiſe the ſtrongeſt wedder will oppreſs the reſt. They muſt be ſhorn when put up to feed, and be always kept clean. There muſt be a heck for holding hay, a place for the turnip or cabbage, and a rip of oats be hung up within their reach. A ſheep, like an ox, eats of turnip the third of its own weight in twenty-four hours. The beſt age for feeding wedders is four or five; ſhort of that age they feed not ſo kindly, nor tallow ſo well.

Some land is fitted for feeding; ſome for breeding only. A farmer who poſſeſſes rich land, buys lean cattle from the breeder, and fattens them for the butcher.

4. The wintering of Cattle not intended for immediate sale.

The food proper for ſuch cattle is, firſt, coarſe graſs that they refuſe to eat during ſummer, when they have better food. Second, ſtraw or coarſe hay; and, laſt, what is left by milk-cows, by beaſts ſtall-fed, or by working oxen and horſes. With reſpect to the firſt, the common but very improper method is, to let them go at large in the

the field; and in froſt and ſnow to give them food in hecks. By this method, the dung is almoſt entirely loſt; and unleſs the field happen to be well ſheltered, which ſeldom is the caſe, Scotland is a climate too cold for ſuch management. This is verified in our highland cattle, which never arrive at half their ſize; not ſo much from want of food, as from piercing cold in a mountainous country. To remedy both theſe evils, an open ſhed ſhould be erected in the field, for ſhelter to the cattle in a ſtorm; and this ſhed ſhould be in the moſt ſheltered ſpot, fronting the ſouth. To preſerve the food given them during froſt and ſnow, a heck ſhould be put in the ſhed; and the dung be carefully collected from time to time. It is evident from what is ſaid, that no cattle are fit to be managed in this manner, but what have been accuſtomed to lie out all winter.

Where there is no coarſe graſs for winter-feeding, the ordinary method to feed winterers is in a dung-yard with ſtraw, the half of which, trodden down with the foot, turns to no account. The more provident farmers ſet up a heck in the yard, which ſaves ſome of the ſtraw that is loſt the other way.

This method, though common, lies open to many objections: firſt, it is ſtill too cold for winter; next, much good food is deſtroyed by it; and, third, the dung, by trampling of the cattle, is is kept from rotting. This matter is interefting, and I beg the reader's attention, while I endeavour to verify thefe propofitions. With refpect to the firft, the fhelter of a yard is far from being fufficient to preferve cattle in a kindly heat, during frofty weather. Such an inclofure is fubject to whirling winds, which pinch the cattle even in frefh weather. Obferve, that when cattle are brought from a warmer climate to mend the breed, every one is fenfible, that during winter a warm houfe, and plenty of food, fhould be provided for them. A cow when thus treated gives more milk, and a beaft intended for the butcher grows fooner fat. To anfwer thefe purpofes, I believe it will be found, that the nearer the air approaches to the heat of the blood, the cattle thrive the better. That cattle fed in a yard deftroy much ftraw, is obvious to ocular infpection. Even when ftraw is put into a heck, one beaft no fooner draws out a mouthful, than he is pufhed away by another, and lofes in his hurry part of what he draws from the heck: none of them are fuffered to feed peaceably or quietly. The third propofition is the moft important of all. Half-rotten dung trodden under foot, and kneaded together by the cattle, excludes the air totally from the inner parts of the heap; and it is a truth indifputable, that there never can be any putrefaction where there is no air: putrefaction cannot go on without moifture, and as little without air.

Put

Put a ſpoonful of cream in an exhauſted receiver; and after a year it will come out as ſweet as when put in. When a dunghill, after heating, turns cool again, turn it over and admit freſh air, it heats a ſecond time: a ſtake driven into an open gravelly ſoil, which admits air into every pore, rots ſooner than even in a wet clay-ſoil that excludes air.

From theſe premiſes I conclude with firmneſs, that the beſt way of feeding winterers is in a houſe where there is free ventilation; indulging them only an hour or two in the field when let out to water, longer or ſhorter according to the weather. I know many judicious farmers, who prefer the feeding on a dunghill, with an an open ſhed to retire to in a ſtorm; and they even hold, that the treading of cattle on a dunghill contributes to its putrefaction. I flatter myſelf that I have diſcovered their reaſon for this opinion. Their practice is, to throw quantities of dry ſtraw from time to time on the dunghill, part of which is eaten by the cattle, and the reſt trodden under foot. Straw is elaſtic, and when laid on a dunghill dry, it admits too much air; in which caſe the dunghill gains by being trodden on. But where all the ſtraw is uſed, either in food or in litter, and none of it carried to the dunghill till it be half putrefied; it becomes too compact by being trodden upon; and the air is excluded altogether, which obſtructs putrefaction

no

no leſs than too much air. A provident farmer will never waſte his ſtraw, by throwing it on a dunghill: he will provide as many winterers, as to conſume all the ſtraw that is to ſpare from his working cattle. I add, that ſuppoſing winterers to be as comfortably put up in a farm-yard as in a houſe, the latter however ought to be preferred; firſt, becauſe it ſaves ſtraw, and the beaſts are more regularly fed; next, that it is a better way of having a dunghill well putrefied; and, laſtly, that the urine of the cattle can be wholly preſerved; whereas in a farm-yard all is loſt but what happens to fall on the dunghill.

Join another particular not of ſlight moment. A dunghill ſeparate from cattle, may contribute greatly to the feeding with turnip. A quantity of turnip taken up in the beginning of a froſt and laid upon the dunghill, will be preſerved entire for weeks, perhaps as long as froſt commonly laſts in this variable climate.

5. Rules for buying and selling Cattle and Corn.

The common and conſtant way is, to ſell cattle by the eye; which is far from being equal between the farmer and the butcher. The former has but a very uncertain gueſs of the weight; the latter, who is dealing the year round, can gueſs very near. Fair commerce is in general the

the moſt beneficial; and every thing that approaches to chance or gaming is hurtful. To bring on a perfect equality between the ſeller and buyer, I propoſe, that every beaſt ſhould be ſold by weight, and that the weight ſhould be aſcertained by a balance. The moſt convenient one I know is a ſteel-yard, being the fitteſt for weighing heavy goods; and a ſteel-yard I have uſed with eaſe and ſucceſs many years.

But it is not ſufficient to have the weight of a living beaſt aſcertained. Different parts are very different in their value; and there is a rule for aſcertaining the proportion of theſe different parts, by which their weight may be known, almoſt with equal certainty as the weight of the whole beaſt. My experience goes no farther than to Scotch cattle ſold fat, to which the following proportions will nearly anſwer. The four quarters make the half of the weight of the bullock. The ſkin is the eighteenth part. The tallow the twelfth part. Theſe make twenty-three thirty-ſixths of the whole; and the head, feet, tripe, blood, *&c.* make the remainder, being a third part and a little more. Theſe offals never ſell by weight, but at a certain proportion of the weight of the beaſt. They commonly draw ten ſhillings and ſixpence when the bullock weighs a hundred ſtone Dutch; and ſo in proportion.

Theſe particulars adjuſted, the next thing the ſeller is to inform himſelf of, is the price of butcher-meat in the market, of tallow, and of hides. Suppoſing the bullock I have to ſell is ſeventy-two ſtone living weight, the four quarters make thirty ſix ſtone, which at 4 s. *per* ſtone, or 3 d. *per* pound, amount to L. 7, 4 s. The hide is worth 16 s. at 4 s. *per* ſtone; and the tallow, being 5 s. 4 d. the ſtone, is worth L. 1, 12 s. Sterling. The offals, according to the proportion above mentioned, will give 7 s. 6 d.; and by that computation the value of the bullock is L. 9, 19 s. 6 d. This anſwers to 2 s. 9$\frac{9}{12}$ d. *per* ſtone living weight. And therefore, if the butcher agree to give me that ſum *per* ſtone, no more is neceſſary to aſcertain the price of the whole cargo, than to weigh the beaſts by threes or fours as the ſcale can hold them. Out of this ſum however muſt be deducted the butcher's profit, which cannot be much leſs than 5 *per cent.*

The weighing cattle alive anſwers another important purpoſe; which is, to diſcover whether the feeder gets the value of the food by the additional weight of the beaſt. For example, the food of a bullock coſts 9 d. *per* day, or 5 s. 3 d. *per* week. If the bullock do not take on two ſtone *per* week, which ſeldom happens, the keeper is a loſer; and there is no excuſe for keeping the beaſt on hand, if it be not the expectation of a riſing market.

There is another advantage of regulating the price of cattle by living weight. Where a butcher

cher buys by the lump, and bargains that he is to take away the beaſt at a day certain, the vender is tempted to be ſparing of food. But ſuppoſing the beaſt to be ſold by living weight, it is the intereſt of the ſeller to make it as fat as poſſible.

The ſame rule in ſelling is applicable to ſheep. The four quarters make half of the living weight, the ſkin the eleventh part, the tallow the tenth part, the offals ſomewhat leſs than the third part. But computing the prices that theſe particulars ſell for in the market, the amount will not be the value of the ſheep, becauſe the wool alſo muſt be taken under conſideration. The wool from ſhearing-time to Lammas is of little value. At the end of December the value is by many underſtood to be half of the value of a full grown fleece; and upon that ſuppoſition, the following calculation is made. Yet there are not wanting expert dealers who value it much higher. They ſay, that in February or March, the wool begins to fork at the ends; after which it increaſes in weight, but little in value. A Linton, Ochill, or Lammermuir wedder, at the age of four, carries a fleece weighing at a medium two pounds eight ounces, valued at 9 d. *per* pound when not tarred, which makes the value of a fleece at clipping-time, 1 s. 10$\frac{6}{12}$d. Sterling. At the end of December it is 11$\frac{9}{12}$ d. *. Thus having a ſcore of

* As the wool of ſheep degenerates into hair in both extremes of heat and cold, ſmearing is a remedy in a cold

of wedders to ſell at the end of December, each of which weighs ſix ſtone fourteen pounds Dutch weight, I want to know what they are worth *per* head.

	L.	s.	d.
1. The four quarters weighing 55 pounds, is, at 3d. *per* pound,	L. 0	13	9
2. The value of the wool as above $11\frac{2}{12}$d. to which muſt be added 10d. as the value of the ſkin, *inde*,	0	1	$9\frac{9}{12}$
3. The tallow being the tenth part of the weight, makes 11 pounds, which, at 4d. *per* pound, is	0	3	8
4. The offals are not ſold by weight, but are commonly valued at -	0	1	0
The ſum,	L. 1	0	$2\frac{9}{12}$

This

cold climate. It alſo kills vermin, and makes a ſheep take on fat five or ſix weeks ſooner than it would do otherwiſe. Some miles round Cheviot hills, none but ſheep of a year old are ſmeared: elder ſheep are reckoned ſufficiently ſtrong to endure the cold without it. But there black ſheep of all ages are ſmeared, becauſe black wool is not diſcoloured by ſmearing. In the north of Scotland ſmearing is not known.

Wool ſhorn immediately after a ſheep is waſhed, never turns white. A kind of oil, termed *gleet*, perſpires from the animal, and contributes to whiten the wool in ſcouring. Therefore, where waſhing is neceſſary, delay ſhearing for a fortnight, that the wool may again be covered with gleet.

This price anſwers to 2 s. 11$\frac{6}{12}$d. Sterling *per* ſtone living weight. And if the butcher agree to give that price, the value of any particular beaſt may be determined by weighing him as above.

From this muſt be deducted the butcher's profit at 5 *per cent.* or under, as can be agreed.

N. B. An animal weighs more or leſs as his belly is more or leſs full. The above proportions were made out when the ſheep and horned cattle were weighed at eleven o'clock forenoon.

HAVING diſcuſſed cattle, I proceed to corn. The variety of meaſures in ſelling and buying corn, has been time out of mind a juſt cauſe of complaint in Scotland, as well as other countries; and it is indeed an inlet to many frauds. And yet, were all the meaſures uſed in Scotland reduced to the Linlithgow ſtandard, there would ſtill remain a great inconvenience in ſelling grain by meaſure; for of all commodities, it is the leaſt proper to be ſold in that manner. The quality of grain differs widely in different ſoils, and even in the ſame ſoil by good or bad culture. One boll of barley weighs eighteen Dutch ſtone, another only fifteen. One boll of oats weighs fifteen ſtone, another only eleven. The ſame difference is obſervable in other grain.

There is another objection againſt ſelling by meaſure. The ſlovenly farmer is ſatisfied with bulk,

bulk, without any regard to the quality of his grain: whereas, were it the rule to ſell by weight, it would make the farmer doubly attentive to the dreſſing of his land, in order to produce the weightieſt grain.

Beſide weight, another circumſtance enters into conſideration for aſcertaining the value of grain; which is, the proportion of the huſk to the kernel.

Middling wheat weighs *per* boll fourteen Dutch ſtone; of which the huſk makes the ſeventh part, or two Dutch ſtone.

Middling barley weighs *per* boll eighteen Dutch ſtone; of which the huſk weighs one ſtone four pounds.

Middling oats weigh *per* boll fourteen Dutch ſtone, of which the huſk weighs ſix ſtone.

Middling beans weigh *per* boll fifteen ſtone eight pounds Dutch weight, of which the huſk weighs but eight pounds,

And the ſame proportion holds as to peaſe.

CHAP. X.

Culture of other plants proper for a farm.

There are many uſeful plants propagated by curious farmers that do not anſwer for food. I confine myſelf to three, the moſt common,

mon, and at the ſame time the moſt uſeful; foreſt-trees, flax, hops. I begin with foreſt-trees, as the moſt complex.

SECT. I.

FOREST-TREES.

CONSIDERING the great quantity of waſte land in Scotland, fit only for bearing trees, and the eaſineſs of tranſporting them by navigable arms of the ſea, one cannot but regret the indolence of our forefathers, who neglected that profitable branch of commerce, and left us to the neceſſity of purchaſing foreign timber for every uſe in life. The Commiſſioners of the annexed eſtates, deeply ſenſible of this neglect, have beſtowed liberally to riſe plantations every where in the King's eſtates. Their laudable example has animated many land-proprietors, to benefit themſelves and their country by the ſame means. The ſpirit of planting is rouſed; and there ſeems little doubt of its ſpreading wider and wider, till this country be provided with timber for its own conſumpt at leaſt, if not for exportation. As an author, I am fond to keep up that ſpirit: and in that view, the preſent chapter on the culture of foreſt-trees, will, I hope, be well received. A regular treatiſe upon the ſubject would require a volume: but as I muſt contract what I have to ſay within the

the nutſhell of a chapter, I ſhall ſtudy brevity, and mention nothing but what appears material.

Trees are propagated by ſeed, by cuttings, by layers, and by ſuckers.

1. Raising Trees by Seed.

The propagating trees by ſeed is nature's method. One inconvenience it has, that the trees thus raiſed are not always the ſame with the parent plant: though they are of the ſame ſpecies, they copy not always its varieties.

What follows will enable us to judge of the maturity of ſeed. Seed incloſed in a *capſula*, in a pod, or in a cone, is ripe when the covering opens by the heat of the ſun. The ſeed of a fruit-tree is ripe when it no longer adheres to the fruit; and where unripe fruit is pulled, the ſeed ripens with it. In general, ſeed is ripe when it ſinks in water to the bottom.

The ſeed of the Scotch elm ripens before the middle of June. The beſt way of gathering it is, to ſhake the tree gently: the ripeſt ſeed falls firſt, which may be gathered in a ſheet laid at the root of the tree.

The ſeed of the aſh and of the maple may be put into the ground without being taken out of its *capſula*.

The beſt way of opening the cones of pine, fir, *&c.* is to expoſe them in boxes to ſun and dew. The

The drying them in a kiln is apt to deſtroy the germ. The cones of the larix are at their full ſize in autumn; but the ſeed is not ſo early ripe. Delay gathering them till March or April, when they begin to drop from the tree. Cut off a part of the cone next the ſtalk, which will render it eaſy to ſeparate the quarters: the ripeſt ſeed falls out upon ſhaking the cone with the hand.

The ſeed of the birch, the willow, the poplar, the aller, being very ſmall, is not eaſily gathered: ſtir the ground about theſe trees, and it will ſoon be filled with young plants. With reſpect to the ſeed of the birch and aſh, it is ſingular, that when dropt from the tree, no ſeed takes root ſo readily; yet when gathered, and ſcattered with the hand, it ſeldom grows.

As for a choice of ſeed, ſmall acorns gathered from large and lofty trees, are preferable before the largeſt acorns of ſmaller trees. In general, the ſeed is always the beſt that is procured from the moſt vigorous trees. But as in extenſive plantations much preciſion cannot be expected, it ought to be the chief care that the ſeed be perfectly ſound.

Next, as to preparing ſeed for ſowing. Trees propagated from ſeed have all of them a tap-root, which puſhes perpendicularly downward. The purpoſe of nature in this root is, to fit trees for growing in the ſtiffeſt ſoil, and to ſecure them againſt wind; but it proves hurtful to trees intended for tranſplantation. A young oak five or ſix years old, when taken up for tranſplanting, has, like a turnip, but this ſingle root, which will be four or five feet long when the ſtem is within one foot. Planted in this manner, it ſeldom lives. This evil is prevented by making the ſeed germinate in moiſt earth, and ſowing it in the ſeed-bed after the radicle is cut off. The radicle never puſhes more; and inſtead of it the tree puſhes out many roots, which ſpread horizontally. Walnuts, almonds, and other ſhell-fruit, being long of germinating, ought to be put in moiſt ſand, in order that the radicle may puſh before the end of April, to be cut off as aforeſaid. Acorns, Cheſnuts, and beech-maſt, will germinate timeouſly in dry ſand. In wet ſand or moiſt earth, they would, before the time of ſowing, not only germinate, but puſh out long roots, which would ruin all. As this method is too troubleſome for ſmall ſeeds, ſow them in beds as gathered: pull them up the ſecond year: cut off the tap-root: and plant them again at the diſtance from each other of three or four inches. Two years after, they may again be tranſplanted wider; there to remain till they be fit for the field. Some imagine, that to deprive a tree of the tap-root prevents its growth. But experience vouches the contrary. And ſo does reaſon; for it is obſervable, that the roots next the ſurface, being acceſſible to ſun and moiſture, are

are always the moſt vigorous, and are farther ſpread than thoſe below. A tap-root is deprived of ſun and air, and even of water, unleſs where it happens to glide below the ſurface: how then can it equal a horizontal root in nouriſhing the tree?

The ſeeds of the white thorn ſown without preparation, riſe not till the ſecond year. If buried under ground in a heap till the pulp be rotted off, and ſown in the ſpring following, they will germinate that very year. Inſtead of burying them under ground, a more approved method is, to lay them in a heap at the end of a barn, mixed with earth. By that method, a greater number will germinate than in the ordinary way. I made an experiment. One bed was ſown with haws prepared in the ordinary way; and one with haws prepared in the other way. Upon the latter bed ſprung a double quantity of thorns, and more vigorous. I made another experiment upon elm-ſeed. Of a quantity gathered when ripe, the half was immediately ſown; the other half was carefully dried in the ſhade, and ſown a fortnight after. The latter produced a greater number of plants, and more vigorous. Thorns are propagated ſtill more expeditiouſly by cuttings from the root. When thorns are taken from the nurſery to be planted in a hedge, the roots that are either wounded by the ſpade, or too long, muſt be cut off. Let theſe be ſhred into ſmall parts, and ſown in a bed

prepared

prepared for them: they will produce thorns that very year. The ſeed of the aſh ſeldom germinates till the ſecond year: when gathered in the month of October, let it be put in pots with earth and ſown in the ſpring: it will germinate immediately. The ordinary way of raiſing hollies, is to ſow the berries entire; which is wrong: every berry contains four ſeeds; and the plants that ſpring from them are ſo interwoven, as not to be ſeparable without injury. A better way is, to gather the berries in December, the later the better, if they can be ſaved from birds. Throw them into a tub with water, and between the hands rub them carefully in the water till all the pulp fall off. The good ſeed will ſink to the bottom, which, after the water is poured off, muſt be laid upon a cloth to dry. Mix them with dry ſand, which will preſerve them all winter. Sow them in March or April, and cover them with earth about three quarters of an inch thick.

With reſpect to the time of ſowing, the beſt rule is, to imitate nature, by ſowing when the ſeed is ripe; provided the tree be of a hardy kind to endure the froſt of winter. By this rule, the ſeed of Scotch elm ought to be ſown in June; the ſeed of pine and fir in April, at which time their cones open. Acorns, cheſnuts, and beech-maſt, ripen in autumn, which is the time of ſowing them. If they ripen later, it is more ſafe to ſow

ſow them in the ſpring following; becauſe the young plants cannot reſiſt froſt, if before winter they have not acquired ſome degree of vigour. There is another reaſon for ſtoring up theſe ſeeds till ſpring; which is, that the longer they lie in the ground, the greater riſk they run of being deſtroyed by vermin. As the white thorn vegetates early, the haws ought to be ſown the firſt dry weather in February, after being ſeparated by a wire-ſieve from the mould with which they were mixed. Avoid freſh dung, which is injurious to them. Sow the ſeed of the larix when taken out of the cone in March or April; for though in the cone it will ſtand good for years, yet it does not long retain its vegetative quality when ſeparated from the cone.

Next, as to the manner of ſowing ſeed. Nature drops ſeed upon the ſurface of the ground. We muſt depart from nature in this inſtance, upon the following account, that after much expence and trouble in procuring ſeed, the far greater part would periſh, partly by vermin, and partly by inclement air. This is not regarded by nature, which is profuſe in the production of ſeed. All ſeeds therefore ought to be covered with earth, birch-ſeed alone excepted, which ought to be preſſed down with the back of the ſpade, but left open to the air without covering. Small ſeeds muſt be ſlightly covered, as having

leſs

lefs vigour to pufh upward *. In ftrong foil, the covering ought in every cafe to be flight. The depth is pretty much arbitrary, becaufe the fame feed will thrive at different depths. But it muft be attended to, that a flight covering expofes the feed to drought; and therefore the ground ought to be watered if the feafon be dry. Where the ground fown is too extenfive for watering, a crop of barley will preferve the tree-feed from the fun, and alfo prevent weeds. The tree-feed and the barley may be fown alternately in lines. If trees are intended to remain where their feed is fown, it is proper to fow thick, partly for fhelter, partly to keep down weeds. M. Buffon declares againft weeding the ground upon which the feed is fown: "For," fays he, "weeds fhelter the "young plants from the fun, keep in the dew, "and preferve the plants warm in winter." In Scotland nothing is more hurtful to plants than weeds, which choke them, and exclude air. A better

* As a flight covering expofes the feed to drought, watering is commonly ufed if the feafon prove dry. But the furface by fuch watering is apt to harden and to prevent the tender plants from fpringing. It is eafy indeed to break the cruft by a harrow or rake; but this lays the plants open to be deftroyed by the fun. A method that has been ufed with fuccefs, is to cover the feed half an inch deeper than ordinary; and from time to time to remove part of the covering. The furface by this means being rendered loofe and free, will be a kindly invitation to the young plants to pufh upward.

better way, even in France, is to ſow barley with the ſeed, which will protect the young plants from the ſun, and admit air.

The beſt way of preſerving ſeed is in dry ſand, which ſucks in the moiſture from the ſeed, and prevents muſtineſs. It withal retains ſo much moiſture as to prevent the ſeed from withering. This method is chiefly uſeful in preſerving during winter ſeeds that require ſpring-ſowing, and in the conveyance of ſeeds to a diſtance. The efficacy of dry ſand appears in preſerving oranges and citrons, which in the air dry and wither: if to prevent withering they be laid in a moiſt place, they never fail to turn muſty. There is one exception, that ſeed which lies long in the ground before it germinates, ought to be preſerved in moiſt earth. The ſeed of the ſenſitive plant will keep entire for twenty years; of a melon for nine or ten. There are many ſeeds that will not keep entire longer than two or three years; which is the caſe of flax-ſeed, though remarkably oily: ſome ſeeds require to be put in the ground as ſoon as ripe.

To prevent young plants in the ſeed-bed from being ſpewed out by froſt, cover the beds with leaves of trees, to be removed when the ſevere froſts are over.

We proceed from the ſeed-bed to the nurſery. Plants form very different roots, according to the ſoil they grow in. In ſtiff ſoil, the roots are commonly

monly few, but ſtrong and vigorous for overcoming the reſiſtance of ſuch a ſoil. Roots multiply in proportion to the richneſs and mellowneſs of a ſoil. An oak, for example, has a ſtrong tap-root, which fits it, more than any other tree, for growing in a ſtiff ſoil. This root diminiſhes in ſtrength and ſize in a loam, and ſtill more in a ſandy ſoil. When it grows in water, it has a multitude of roots, but not the leaſt appearance of a tap-root. Hence it follows, that the ſoil of a nurſery ought always to be light and free: ſuch a ſoil produces a multitude of roots; and the vigour of growth is always in proportion to the number of roots, the ſmaller the better. But it alſo follows, that in tranſplanting trees from ſuch a nurſery, the ſoil about them ought to be made as mellow and free as poſſible, in order to encourage the ſmall roots. When theſe are enlarged in ſo fine a ſoil, they will be able to overcome the ſtiffneſs of the natural ſoil of the field. Avoid dung in a nurſery. If any be admitted, it ought to be thoroughly putrefied, and digeſted into a ſort of rich mould. Green dung makes the roots ill conditioned, and encourages a large white worm, which lives on the bark of the roots. Neither the walnut nor horſe-cheſnut ſucceed in a nurſery: the plants require to be placed at a diſtance from each other; and the earth about them muſt be ſtirred ſeveral years. Aquatics that are intended to be propagated by

large

large cuttings, ought firſt to have the benefit of a nurſery; becauſe they thrive beſt when planted out with roots. Avoid a mixture of different trees in the ſame bed, for the ſlow growers will be oppreſſed.

The true ſeaſon for tranſplanting from the ſeed-bed to the nurſery is about the fall of the leaf. Catch the time when the earth is ſo moiſt as to ſuffer the plants to be drawn without tearing the roots. All evergreens ought to be tranſplanted in ſpring; and alſo all other trees that ſuffer by froſt.

Where trees are ſo young as that an interval of five or ſix inches along the rows is ſufficient, there muſt be an interval of a foot at leaſt between the rows, in order to give acceſs to clean the ground of weeds; and this interval is ſufficient, even when the plants are ſo large as to make an interval of a foot along the rows neceſſary. Where the diſtance along the rows is made eighteen inches, or two feet, the intervals between the rows ought to be no leſs, for the ſake of the trees, though unneceſſary for the ſake of weeding. Yet ſuch is the influence of cuſtom, contrary to common ſenſe, that from the original poſition of young plants in a nurſery, the interval between the rows is always made double of the interval along the rows. Thus if the latter be eighteen inches, the former is always made three feet; and four feet where the ſize of the trees requires an interval of two feet along the rows. The ſame influence of cuſtom occaſions trees to be planted in rows in the field, where they are to ſtand; and yet they make a much better figure when, in imitation of nature, they are ſcattered as at random.

The pruning young trees in the nurſery is an article that deſerves attention. Lateral branches that are like to get the better of the ſtem, ought to be retarded in their growth by being pinched more than once during ſummer: and where the lateral branches are too frequent, they ought to be thinned. This practice promotes greatly the growth of the ſtem. The like good effect will follow from treating in the ſame manner, for two or three years, young trees after they are planted out. Thus managed, they will for ever after need very little pruning.

2. Cuttings, Layers, Suckers.

Seed is not the only means that nature has provided for propagating trees. They can be propagated by cuttings, by layers, and by ſuckers. They have one advantage, that whereas ſeed propagates the kind only, theſe carry on any variety in the tree from whence they are taken. Grafting has the ſame effect.

The willow, the oſier, the vine, the elder, the poplar both black and white, the platanus, the yew, the box, may all be propagated by cuttings.

The

The black poplar is ſingular: a branch will not anſwer for a cutting, unleſs the top be left entire. A cutting of willow eight or ten feet long, and ſeven or eight inches thick, planted in the following manner, will ſoon become a tree. Let the great end be immerſed in water a foot deep as ſoon as cut, which ought to be the end of March; and remain in that ſtate till planted, which ought to be the beginning of May. Immediately before planting, cut the great end ſloping to a point; and on the ſide that is not cut, keep the bark entire down to the point. Make a hole in the earth a foot and a half deep, ſufficiently large to admit the cutting without ruffling the bark, which would prove deſtructive. Any vacuity left in the hole muſt be filled up with freſh earth, and preſſed down cloſe to the cutting. A ſafer way, and little leſs expeditious, is to make the hole with the ſpade, to return the earth to the hole round the cutting, and to preſs down all firm and cloſe. After ſtanding a year, throw ſome freſh earth upon the roots. It is ſtill better, to make a ditch two feet from the row of the trees thus planted, and to lay upon the roots the earth of the ditch. Several trees that can be propagated by cuttings, require more precaution than the willow or poplar. The branches intended for cuttings muſt be young; and a part of the greater branch from whence they are taken muſt be cut with them, to ſerve as a ſort of root. With all

theſe

these precautions, it must not be expected, that every cutting from the platanus, the white poplar, the poplars of Virginia and of Lombardy, the aspin, the maple with ash leaves, will take root. Cuttings from the yew, the alaternus, the box, require the utmost care. They ought to be planted in beds sheltered from the sun, and watered in dry weather.

All trees do by layers; and the method is easy in shrubs which have branches near the ground: but to propagate by layers from large trees that have no branches near the ground, such as the lime, the mulberry, the aller, more art is necessary. One method is, to cut the tree by the ground before the sap begins to rise; which will produce many shoots the first year. The next year at the same time, heap earth upon the root till it rise a foot upon the shoots. After two or three years remove the earth, and cut off the shoots, which will be full of roots as far as they were covered with earth. These will make good plants. Spare the smaller shoots, and cover them with earth as before: they will in time produce new plants. By the following method, the shoots may be so managed as to produce young plants without end. Toward the end of February, bend the best shoots down to the ground, twisting them a good deal about the middle, and covering that part with earth. Set the ends upright to be preserved in that posture by earth pressed about them.

them. A ſhoot by being twiſted, will produce more roots the firſt year, than otherways in two or three. Every ſhoot thus laid, will produce ſeveral ſhoots; of which layers may be made, managed as above-mentioned. A third method is, to chuſe a young tree eight or nine inches in circumference; and cut it over two feet from the ground. In every part, the ſtump puſhes many ſhoots. The ſecond or third year, take all out of the ground: lay the ſtump with its roots and ſhoots in a trench; and cover them with earth, ſo as to leave nothing open to the air but the extremities of the ſhoots. Theſe will thruſt out roots under the ſurface, which will afford young plants in abundance. When a pine or a fir is cut down, the root dies without producing any ſhoot. Nor can theſe trees be propagated either by cuttings or by layers; if a ſpruce fir be excepted, which it is ſaid can be propagated by cuttings.

To have ſtraight trees, cuttings and layers ought to be of perpendicular branches; eſpecially where the wood is hard. How long layers ſhould continue in the ground to have good roots, depends much upon the ſeaſon, and ſtill more on the nature of the tree. A bramble will take root on the ſurface of the ground, without the leaſt covering of earth. Layers of lime and of platanus have commonly good roots in three years, and

and ſometimes in two. Many trees require longer time.

To propagate by ſuckers, part of the root of the parent plant muſt be cut along with the ſucker. It muſt be immediately planted in a nurſery, in order to acquire roots. When planted out to ſtand, the old root ſhould be cut away all but what is cloſe to the ſucker; for it is apt to ſwell under ground, and to obſtruct the growth of the plant. To have a quantity of ſuckers, which riſe not plentifully from old trees, cut the old tree over, and the ground will be filled with them. Some writers ſay, that trees propagated by layers take on a better ſhape, and grow faſter, than thoſe propagated by ſuckers.

3. Soil proper for Trees.

The ſoil proper for trees comes next in order. Fat ſand, that is, clay mixed with a large proportion of ſand, agrees with almoſt every ſort of tree, eſpecially if the ſoil be deep. Evergreens do well in ſuch a ſoil. And even aquatics, ſuch as aſh, poplar, willow, aller, thrive in it, though more ſlowly than in moiſt ground: theſe aquatics make a ſhift even in ground too dry for the oak, the beech, the cheſnut. A pine grows well in ſand. The juniper will grow in a very thin ſoil where ſcarce any other tree will thrive. An aſpin does well in a pure clay: a cheſnut does not. Dry earth

earth of a good quality, though but eighteen inches above a tough till, will bear the elm, the maple, the horn-beam, the walnut, the aſh, the birch, the mulberry, the white poplar, and almoſt every kind of ſhrub. If but ten or twelve inches deep, it is only fit for ſhrubs ; excepting the birch, which will grow if the till be covered with but five or ſix inches of black and light earth. The lime and the horſe-cheſnut love a tender and deep ſoil. Marſhy ſoil is proper for the willow, the aſh, the aller, the occidental platanus, and for moſt ſorts of the poplar. An elder, though it makes a ſhift in the drieſt ſoil, yet loves moiſture ; and therefore is proper for a fence on the ſides of ditches, for cattle do not touch it. Ground elevated two or three feet above a running water, being moiſt without being marſhy, is fit for trees of every ſort, particularly for the occidental and oriental platanus, the tulip-tree, and the lime. Young trees ſhould be planted differently, according to the nature of the ſoil : in a ſoil retentive of moiſture, the tree ſhould be bulked above the ſurface ; in a very dry ſoil, the earth round the tree ſhould ſlope down to the root, in order to collect the rain. In general, though corn proſpers in clay, ſandy or gravelly ſoil is fitteſt for trees. The culture of corn, renewed annually, renders clay open and free ; but when left untilled, it turns too hard even for the roots of trees.

4. CLI-

4. Climate.

With respect to climate, an oak is an inhabitant of the temperate zone: none are found between the tropics, and none farther north than Stockholm. Fir and birch bear much cold, the latter especially. Plants, like animals, after several generations, come to thrive in a climate very different from their own. It requires peculiar skill and attention to habituate to a new climate such as grow on the top of the Alps, and on the top of high mountains farther north. Their nature fits them for a very cold summer; and for a temperate winter, being always covered with snow during that season. In a low country the want of snow can be tolerably well supplied by some small degree of artificial heat; but in our summers it is not easy to give them the cold of their native place.

The advantages and disadvantages of different exposures come under the present article. In spring, even a strong frost hurts not plants, provided the ice be melted before they are exposed to the sun: but even a moderate frost commits great waste, if the ice continue till it be melted by the sun; for the ice acts as a burning-glass. By that means, delicate plants, exposed to the rising sun, are often destroyed in spring; while such plants, exposed to the north, are safe: which is the

the fate even of young ſhoots of the oak. An eaſtern expoſition has, on the other hand, its advantages. Plants are ſooner relieved from the morning cold by the heat of the ſun : they are protected from the ſun's meridian heat : and as the eaſt wind is generally dry, ſpring-froſt makes leſs impreſſion than where there is more moiſture.

In a ſouthern expoſition, plants, being covered from the north, are the leſs ſubject to froſt. As the ſun reaches them not till about ten in the morning, the ſpring ice is commonly melted before that time ; and if the climate be rainy to temper the meridian heat, trees in that expoſure grow faſt. The diſadvantages are, firſt, that young plants are apt to be ſcorched by the ſun, eſpecially in a light ſoil and dry ſummer : next, that ice is frequently melted in that expoſure even by winter-heat ; and if the moiſture be not ſucked up by heat or diſſipated by wind, it congeals in the evening, and does much hurt.

A weſtern expoſition is free from the burning ſun in a dry ſeaſon ; but is expoſed to ſpring ſnow and hail, which come moſtly from the weſt. We often feel a weſt expoſition inſupportably cold, when the air is tolerably ſoft in other expoſitions. Hardy trees can bear this cold ; but tender trees are often deſtroyed by it.

In a northern expoſition, ſnow never melts ; and the wind from that quarter is the coldeſt and drieſt of any. By that means however perſpiration is ſo faint, that little moiſture is required. Here deciduous foreſt-trees grow ſlowly; but the pine, the fir, the yew, the evergreen oak, the box, and the other evergreens, thrive well. This fact appears ſingular: evergreens perſpire little; and one would think that they ſhould, more than other trees, need the action of the ſun to keep their ſap in motion. The birch alſo thrives well in a northern aſpect. In a dry and light ſoil, trees thrive better expoſed to the north, than to the ſouth: in the former, the dew continues on the trees till nine or ten in the morning; and the ſun ſtrikes not ſo violently as to wither the ground, or to occaſion great perſpiration. But a ſouthern aſpect merits the preference in a clay ſoil, and in a cold climate.

5. Time of planting trees in the field.

With reſpect to the time of planting trees where they are to ſtand, it runs from autumn when they drop the leaf, till ſpring when the buds begin to ſwell; provided it be freſh weather, and the ground be tolerably dry. But becauſe in the dead of winter froſt prevails, or too much moiſture, two ſeaſons chiefly are recommended, namely autumn before ſtrong froſt ſets in, and ſpring after it is paſt. In this ſeaſon, planting may be continued till the buds begin to open; which

which in ſome trees is early, in others late. The opening the buds depends alſo on the ſeaſon, late or early. Whether autumn or ſpring be the beſt for planting is not agreed. As the weather is commonly moiſt in autumn, trees brought from a diſtance ſuffer leſs in that ſeaſon by being long kept out of the ground; and if the winter happen to be mild, new roots are produced, which are a fine preparation for a more vigorous vegetation in ſpring. Millar holds October to be the beſt ſeaſon for planting trees that loſe the leaf in winter, provided the ſoil be dry. In a moiſt ſoil, it is better, ſays he, to defer planting till the latter end of February or beginning of March. But though this may hold in general, there are ſeveral exceptions. Tender plants that are hurt by froſt, ought to be delayed till ſpring. Trees that retain the leaf during winter, ought to be tranſplanted in ſpring; for, having a ſlow circulation of ſap, they perſpire little, and run leſs riſk of being hurt by drought. Another reaſon concurs with reſpect to the oak, the beech, and the hornbeam, which hold the leaf all winter: the reaſon is, that they are late in puſhing; and when planted in February or March, they have time to prepare new roots before the buds begin to ſwell *. April is undoubtedly the beſt ſeaſon

* It is maintained by ſome writers, that the beſt ſeaſon for planting oak and larche, is immediately before the buds begin to puſh. I have not diſcovered any good reaſon for making theſe trees an exception from the general rule.

ſeaſon for evergreens; though they may be ſafely removed at Midſummer, if near the place where they are to ſtand. During winter evergreens are in a ſtate approaching to reſt. If removed at that ſeaſon, they do not take root till the ſpring ſets the ſap in motion; and in a hard winter they commonly die. Light land ſhould infallibly be planted in autumn; for ſpring-planting in ſuch land is ſubject to great diſtreſs from the ſummer-ſun.

6. Manner of Planting.

As to the manner of planting trees in a field, the ſafeſt way to draw them out of a nurſery for planting, is to remove the earth from the firſt row of trees, which by that means will be eaſily drawn without hurting the roots; proceeding in the ſame manner till all be taken up. What are too ſmall, may be planted in a new bed prepared for them. For three years after trees are tranſplanted from the nurſery, the ground about them ought to be twice ſtirred annually. Once a year after will be ſufficient, till the trees get entirely the better of weeds. Plantations weeded for a year or two only, have viſibly languiſhed. As ſoon as trees thus cultivated begin to join their branches, the lower branches fall off, and the weeds are ſmothered. This is the time of the moſt vigorous growth. Trees cannot be planted too thick, provided attention be given to thin

them

them timeoufly. Thick planting fhelters them from wind and from cold. When they rife to feven or eight feet height, fo as to wave with wind, they ought to be thinned, to prevent their rubbing upon one another, which is apt to create a fort of gangrene. But let them be thinned cautioufly: fhorten the branches of the weakeft tree, for giving room to the trees that are to ftand; and let that tree be cut down a year or two after. This is a better method than to cut down the tree at firft; for young trees that have too much room, are apt to branch out inftead of pufhing up. This method does finely in a plantation of firs and pines. When the trees rife to eighteen or twenty feet, and by good roots are proof againft wind, they ought to be much thinned, fo as not to leave one nearer another than its own length. Where trees of that height ftand clofe together, they occafion a ftagnation of air: the air they fuck in is unwholefome, and makes them languifh. Young trees five or fix feet high only, occafion very little ftagnation of air: the free air above mixing with that below, preferves all fweet and wholefome.

Free air is neceffary for plants as well as for animals. People crouded together in a great town, lofe vigour and become difeafed. For the fame reafon, trees planted thick, wither and lofe their growth. Of this, their are examples without number in Angus and Mearns. We fee every

every where clumps of firs, many that have ſtood thirty or forty years, without lateral branches, not a leaf but at the top where the air is free; the outer rows where the air can penetrate, tolerable; of the reſt of the clump, within, not one tree bigger than a man's leg. Yet, in no other part of the world can there be a ſtronger inducement to thin trees; not only to procure timber, which is much wanted in that country, but to procure a ſtill more uſeful commodity, and that is fewel; which is ſo ſcanty in that populous and manufacturing country, that for an acre of broom of five years growth, L. 5 Sterling is an ordinary price. Yet the proprietors in other reſpects, are careful of their affairs. It is not in my power to find a meaning. Suppoſe a field of fine wheat is ſuffered to rot on the ground, without ever applying the ſickle; would it not be juſtly concluded, that the farmer is crazy? Is there no reaſon to apprehend the ſame imputation upon gentlemen, who, after the expence of fencing and planting, look on and ſuffer their plantations to go to ruin, by neglecting to apply the axe? I have often cenſured this ſupine negligence, as not only hurtful to the proprietors, but to their manufactures. I willingly embrace this opportunity of public admonition, hoping it may prove more ſucceſsful than private cenſure.

One method of making a plantation turn to profit in a ſhort time, is to begin with planting birch

birch in rows with intervals of ſix feet, ſtirring the ground about them two years. In the intervals ſow acorns, cheſnuts, or beech-maſt. The birch, a faſt grower, will in a few years form a thick wood, deſtroy the weeds, and ſhade the young plants below them. The birch cut out when they begin to oppreſs the trees under them, will afford ſome profit, and leave the ground plentifully ſtored with better trees.

The beſt way to encourage trees planted in a row, is to draw a trench at the ſide of the row, at ſuch a diſtance as not to wound the roots. The earth thrown upon the roots will ſave weeding, and ſecure the trees againſt wind. The water in that trench will be an additional nouriſhment to the trees.

Evergreens, perſpiring leſs than other plants, require leſs nouriſhment, and for that reaſon continue green all winter. They reſemble the exanguous tribe of animals, the frog, the toad, the tortoiſe, the ſerpent, which, perſpiring little, make a ſhift to paſs the winter without food. Evergreens, by perſpiring little, have a thick, viſcid, oily ſap, which enables them to endure the winter's cold. They ſeem many of them to thrive beſt in the temperate ſeaſons of the year; not ſo well in the heat of ſummer, their perſpiration being then too great for the ſlow aſcent of the ſap. Evergreens, when tranſplanted, do not ſo readily ſtrike root as other trees: for which

which reaſon they ought to be taken up with a bulk of earth about their roots, Gardeners commonly place ſhrubs of this kind in an oſier-baſket, which ſoon rots after it is put into the ground with a tree. A holly thus tranſplanted ought to be watered in the month of May, if the ſeaſon be dry.

In order to tranſplant a tree with a bulk of earth about the roots, cut a trench round it at the diſtance of ten or twelve inches, as deep as the roots go, cutting over with a ſharp knife all the roots that appear. Return the earth into the trench. Repeat the operation next year, but a little farther from the tree. When the tree is tranſplanted the year after, the roots will be ſo interlaced in the bulk, that none of the earth will fall away in carrying. The hole where the tree is to ſtand, ſhould be conſiderably larger than the bulk, and made ſome months before planting, to receive the influence of ſun, rain, and froſt. Fill up the hole with freſh good earth, fit to encourage the young roots. The harder the ſoil is, the hole ought to be the larger, to receive the greater quantity of the nouriſhing earth. In making the hole, lay the upper *ſtratum* on the one ſide, and what is below on the other: lay the former about the roots, being the beſt ſoil, and cover it with the latter. Preſerve as many roots as poſſible, ſhortening thoſe only that are too long, and what are torn in taking up. If the

the tree have too much head, ſhorten the largeſt branches, leaving the ſmall branches for carrying leaves: it is by theſe that the ſap is drawn up; and it is by theſe that perſpiration is performed. But preſerve carefully the leading ſhoot or ſhoots; for many experiments have made it evident, that theſe have the greateſt power to draw up the ſap. If the lime be not an exception, a tree, after its head is lopped off, ſpreads into a buſh, but never riſes in height. Conclude all, by preſſing down the earth vigorouſly with the foot. A tree planted deep in the ground, ſtands the firmer againſt wind: its roots are better protected againſt ſun and froſt, and leſs apt to throw up ſuckers. Theſe conſiderations notwithſtanding, trees ought not to be planted deep; for they languiſh till they acquire roots near the ſurface. Theſe, ſpreading themſelves in the beſt ſoil, and being acceſſible to ſun and rain, convey much more nouriſhment to the tree than thoſe below. Nothing however is meant againſt planting large trees deeper than what are ſmall. At the ſame time, if the under ſoil be bad, or very moiſt, even large trees ought to be planted near the ſurface, with a bulk of earth about them. Becauſe in a porous ſoil the ſun penetrates deeper than in clay, and withers the roots near the ſurface; for that reaſon a tree ought to be planted deeper in the former than in the latter. Many think, that a fir does beſt where fir-trees have been recently cut:

a fir being the native of a cold country, ſtands much in need of a ſhade in ſummer; and therefore ſucceeds the beſt when protected from the ſun by ſurrounding firs. So far the obſervation holds, and I believe no farther.

It ought to be an indiſpenſable duty, to viſit young plantations after every high wind, and to ſet upright thoſe that are ſhaken, preſſing the earth cloſe about the roots. Support a tree much ſhaken with a forked ſtick. I know no general rule more important than this. It is my opinion, that more trees are loſt by neglect of this operation, than any other way. The operation ought to be renewed after every high wind, till trees have acquired ſuch roots as to ſtand firm againſt wind.

Some are careful, to give a tranſplanted tree the ſame poſition with reſpect to the ſun, that it had in the nurſery. But experience has proved this precaution to be uſeleſs.

As the expence and riſk of filling up a vacant ſpot with a full-grown tree is great, a nurſery may be ſet apart for rearing young trees till they be from fifteen to eighteen feet high. Theſe trees may be tranſplanted twice or thrice in the nurſery for multiplying their roots; and they ought to be planted where they are to ſtand, with as much as poſſible of the nurſery-ſoil about the roots. In that view, froſt is the proper time for the operation. Such trees will not at firſt make

ſo

ſo great a figure as full-grown trees; but there will be leſs hazard of miſcarrying.

Where trees of unequal growth are planted together, the quickeſt growers will overtop the reſt, and deſtroy them. Therefore let trees of the ſame kind be planted in clumps; or let none be mixed but what grow equally faſt. I imagine there is a beauty in ſmall clumps, as it makes different ſhades the more conſpicuous. Colours, even the fineſt, make but a faint impreſſion in a confuſed mixture.

7. PRUNING.

AND now of pruning. By multiplied experiments it is aſcertained, that the cutting off young branches, hurts not any ſort of tree, not even the walnut, nor the reſinous kind. The loweſt branches, which in time would wither and drop off, ought firſt to be cut; but gradually, that the tree may not be bared of leaves. Severe pruning makes trees grow tall, but withal ſo ſlender as not to reſiſt wind. Every tree ſuffers by having its large branches cut off, but in different degrees. The wound of an elm is ſoon healed: not of a walnut, an oak, or a pine; if the tree be not vigorous, the wound is apt to rot into the body. When a large branch is cut off, though the place of its inſertion is ſoon covered with wood and bark; yet the new wood doth not unite with the

the old, and there always remains a defect within. When an overgrown branch is cut close to the trunk, which ought to be at the end of the year, the young shoots must the next June or July be pulled off with the hand. If cut off with a knife, shoots will grow without end, and create a disfiguring bunch in the tree. A writer, whose integrity equals his experience, affirms, that a branch cut some inches from the trunk will make no blemish in the tree *. The greatest attention ought to be given to the head: if forked, one of the rival branches must be cut off six inches high; and to the remaining stump ought to be tied the other branch, in order to give it an upright direction. If a young tree become stationary, as it is apt to do in stiff soil, cut it over by the ground in September, or rather below the ground. Next June or July tear off with the hand all the shoots but the strongest, and lay earth round the stump. If this be delayed till the sap fall, there is danger of tearing the bark of the stump, which would hurt it. Avoid a knife; for at every amputation shoots will push, which must again be cut, and a bunch will grow. Instead of cutting by the ground a young tree become stationary, some recommend a slit in the bark from the root upward. A trial was made of these two methods in the same field; and those that

* Kennedy.

that were ſlit became the moſt thriving trees. The experiment was made upon aſhes.

Pruning may go on from the beginning of July till the middle of September: trees later pruned ſuffer by cold. A holly is an exception: it has a large heart, and is hurt by froſt if clipped or pruned after the month of July. A holly-hedge, that was not clipped after the month of June, endured without hurt the hard froſt of the year 1740.

The Scotch fir, which is not a fir but a pine, deſerves to be cultivated for beauty as well as for uſe. In a natural wood, the timber is always better, and freer of knots, than in a plantation. Trees are always planted at the ſame time; but in a natural wood are of different ages, the elder bearing down the younger, but not without loſing all their under branches. This in effect proves a benefit; for as the under branches do not ſpring again, there happens to be much timber without knots. In a planted field, on the contrary, the trees grow all equally; and if weeding be neglected they grow up like ſpindles without a lateral branch: if duly weeded, ſo as to give room for lateral branches, the lower branches die by degrees, and leave knots in the trees. Therefore to train a fir-plantation, the trees ought to be thinned as ſoon as they begin to wind-wave; and the loweſt tire of branches ſhould be cut off from every tree that is intended to ſtand. Another tire ſhould be cut off the next

next year; and ſo on, leaving five or ſix feet of the trunk without a branch. All the wood that is added to the tree after, will be ſolid without a knot. A better way is, to tear off with the hand the under branches, which bring their roots with them: the hollow is ſoon filled up with regular fibres, and the wood is all equally good. A branch cut away with a knife, though cloſe to the ſtem, leaves its root, which becomes a knot when ſurrounded with new wood. If a Scotch fir be intended for beauty, let all other trees be kept at a diſtance, and it will ſtand forty or fifty years without loſing a branch. Travelling in Northumberland, Durham, Yorkſhire, clumps of Scotch fir are ſeen advantageouſly planted on knolls. But never an attempt to weed. They are in the high way to deſtruction for want of air and room. When a larche is fifteen or ſixteen feet high, cut off the loweſt tire of branches in October cloſe to the ſtem, and the next two tire ſucceſſively, year after year. The rays of the ſun will be admitted to the root of the tree; and the air will have a free circulation, neceſſary to all plants. It will not be neceſſary afterward to uſe a knife to a larche, unleſs it be to lop off a broken branch *.

Sheep

* In Mr Duff's garden, town of Ayr, there was an old ſtump of a thorn three or four feet high, that appeared to have very little life in it. It was pierced in ſeveral places with an iron nail: buds came out at every hole; and now it is a flouriſhing tree about fifteen feet high.

Sheep and goats are more deſtructive to young trees than horned cattle; and theſe, more than horſes. Sheep are fond of the leaves of the white thorn, above all other food: it is folly to think of white-thorn hedges, without baniſhing them. The depredations of theſe creatures upon my plantations, have often ſwelled my heart againſt Providence. Why has it made young trees ſo palatable to them: what woods might not be raiſed, were it not for the expence of fencing! Such thoughts were apt to ſteal in upon me. But in nothing are we more in the wrong than in murmuring againſt Providence. It is to us a ſignal bleſſing that domeſticated animals feed on young trees: otherwiſe, the utmoſt induſtry of man would be inſufficient to prevent trees from occupying the whole ground, and putting an end to agriculture; he would be reduced to his original ſavage ſtate.

8. WOOD PROPER FOR INSTRUMENTS OF HUSBANDRY.

AMONG the moſt expenſive articles of a farm, are carts, ploughs, brakes, harrows, rollers, *&c.*: and it is of importance with reſpect to œconomy, that theſe ſhould be conſtructed of proper wood. To that end the following hints may be of uſe.

I begin with examining, at what age a tree is in perfection for the purpoſes of a farm. At the age

age of ſixty, it is ſufficiently large for every farm-purpoſe; being, when cut to the ſquare, from twelve to fifteen inches each ſide. I muſt except the oak, which, even for the purpoſes of farming, improves till it be a hundred years old. Every oak conſiſts of red and white wood; the former the firmeſt of all wood, the latter good for nothing. Aſh, after the growth of ſixty years, turns ſhort and brittle.

The proper ſeaſon for cutting a tree is, when it has leaſt ſap; which is preciſely in the middle between the time of ſhedding the leaf and that of budding: in that interval it is tough, and fitteſt for every farm-purpoſe. When cut in the ſap, the wood is ſhort, and apt to ſplit with drought.

For preſerving wood after being cut, there are three methods. One is, to dry it in the air; another, to immerſe it in water; and a third, to cover it with horſe-dung. Aſh, when ſawed green, never fails to ſplit. Before applying an inſtrument, it ought to be expoſed eighteen months in a dry ſituation, that all the ſap may evaporate. During that time, both ends ought to be covered from the air: the bark prevents the body from ſpliting; but when the ends are expoſed, they will ſplit into the body five or ſix inches. When aſh is deſigned for uſes that require ſplitting, let it be ſplit immediately after cutting, and the parts laid up where the air has

not

not free acceſs, in order that they may dry by ſlow degrees; for ſudden drought makes them warp. Oak and elm require the ſame treatment. The Huntington willow, and other willows that riſe to a large ſize, turn extremely tough when dry; and therefore, if intended for planks or boards, they ought to be ſawed directly after being cut. But as in this caſe they are apt to ſplit, great care ought to be taken to dry them ſlowly. Alder and birch ought to be managed in the ſame manner.

The immerſing in water, and covering with horſe-dung, are far from being the beſt methods of drying wood. It is always harder and tougher when dried ſlowly in the air. Therefore theſe methods are only for expedition, in order to extract the ſap the more quickly when the wood is wanted for immediate uſe.

There is not an inſtrument of huſbandry that conſiſts of different parts, but requires wood of different kinds. Of all wood, oak is that which reſiſts moiſture the beſt, and can the beſt endure the being totally deprived of air. For theſe reaſons, oak is the only wood fit for being mortiſed into other wood. From the part that is mortiſed, air is totally excluded; and yet ſome moiſture finds acceſs, being more penetrating than air. Therefore, the ſpokes of a wheel, which are mortiſed both into the nave and fillies, ought indiſpenſably to be of oak; the ſheths of harrows, which bind the parts together, ought to be of the ſame wood ; as alſo the head of the chain-plough, becauſe it is mortiſed into the beam. As aſh is leſs apt to ſplit than oak, it is more proper for naves and fillies. Being the tougheſt of all wood, and the moſt elaſtic, it is the fitteſt for the ſhafts of a cart. The beſt wood for the body of a cart or waggon, is the Huntington willow. It is both lighter and tougher than even the beſt red fir. The head of the Scotch plough may be of alder, becauſe it is not mortiſed into any other part. Whatever the plough be, the mouldboard may be of willow, or alder, or plane ; becauſe they are light, and not apt to ſplit. The bulls of brakes and harrows ſhould be of birch or alder. A roller ſhould be made of beech-wood, being heavy ; the ſheths and pins of oak, and the ſhafts of aſh. Foreign fir is the beſt and cheapeſt for couples. Beech-wood would be ſtill better, were it not apt to take the worm : but in a farm-houſe that is not lofted, will not the japanning with ſmoke prevent that evil ? The handles of ſpades, ſhovels, picks, *&c.* ought undoubtedly to be of aſh : beſide its toughneſs, it is leſs apt to turn warm in handling, than any other wood. For gates, fir is undoubtedly the beſt : It is light ; it reſiſts moiſture ; and is not apt to warp.

One general rule I give, of more importance than at firſt view may be thought, which is, that the angle made by mortiſing, or otherwiſe, being always

always the weakeſt part of the inſtrument, ought to be fortified with a plate of iron fitting accurately the angle of the wood.

SECT. II.

FLAX.

HAVING finiſhed what I had to ſay on foreſt-trees, flax comes next in order, according to the diſtribution made above. The regulations publiſhed by the truſtees for manufactures, have left me little to obſerve upon that article. I ſhall venture a few particulars only, more immediately connected with huſbandry. Annual weeds abound ſo much in this country by careleſs management, that the weeding of flax is the greateſt incumbrance on its culture. After horſe-hoed turnip, cabbage, or potatoes, flax ſucceeds well; and I know no other crops that extirpate weeds ſo effectually. A potato-crop being removed in October, the ground during winter gets a froſt-preparation; which cannot be obtained after cabbage or turnip, becauſe theſe keep the ground all winter. Therefore, above all, I recommend a potato-crop as a preparation for flax. Next to it, if at all inferior, is paſture-ground three or four years old, left rough in foggage, and in that ſtate trench-plowed before winter. It gets the froſt-preparation, is clean of weeds; and the rough foggage,

foggage, turned to the bottom of the furrow, retains moisture like a spunge, which it yields plentifully to the roots of the flax. Flax is a thirsty plant; and a better situation it cannot have for procuring nourishment.

In pulling flax, the fully ripe ought to be separated from what is less so, and the tall from the short: when mixed, they neither water well, nor dress well. Let the seed, when separated from the plant, be beat lightly and passed through a sieve. The best and plumpest will come out first; and this ought to be reserved for sowing again. Beat what remains with a heavier hand: the seed thus got will be fit for the oil-mill.

Lint pulled green requires less watering than when fully ripe.

SECT. III.

Hops.

THE last article of this kind I undertook, is the hop. A regular hop-yard is an undertaking too great for an ordinary farmer; but every farmer, by the following plan, may have hops of his own growing, sufficient for his own use, and perhaps for his neighbours. If I can reduce the expence within moderate bounds, I am not afraid of the climate of Scotland: many judicious

judicious trials made here, have produced good hops.

Dr Woodward has obſerved, that the fruit which grows neareſt the ground, is always the beſt. This has ſuggeſted to me a thought, that hops may be propagated in the eſpalier way like apples or pears. Fix in the ground, at an interval of three feet, a number of poles eight or nine feet high, in a line from weſt to eaſt. Beginning at the weſt end, plant a hop-vine at the foot of each pole, the laſt ſix or ſeven excepted. Inſtead of allowing them to aſcend the poles, train them from weſt to eaſt in an angle with the horizon of nine or ten degrees, and directing them from pole to pole by ſmall twigs between the poles; obſerving to twiſt them round the poles and twigs as they grow naturally in aſcending upon a pole: in the contrary direction they cannot be made to grow. In this manner, each vine may extend itſelf twenty or thirty feet, without riſing at its extremity more than nine feet above the ground. If a hop reſemble other fruit-trees, it will carry more fruit by having the growth of the wood checked. And if this method ſucceed, a ſmall hop-plantation may be within the reach of every farmer, requiring ſome attention, but little expence. Poles above twenty feet long, renewed every two or three years, are a moſt expenſive article in an Engliſh hop-yard; and the places in Scotland are few where they can be procured at

any

any rate: but poles of nine or ten feet may be procured every where. There is another ſignal advantage of this method: Wind is a great enemy to a hop-yard; but a row of humble poles bound together by hop-vines will be ſufficiently ſecure againſt wind, eſpecially in a line from weſt to eaſt.

Taking it for granted, that the fruit next the ground is the beſt, I am in the courſe of an experiment with a young apple-tree having two branches, one trained to the weſt, the other to the eaſt, upon ſmall pegs of wood about ſix inches high. Under theſe branches the earth is covered with ſharp ſand, in order that the fruit may be benefited both by the direct and reflected rays of the ſun. It deſerves an experiment, whether hops may not be trained ſucceſsfully the ſame way, without the expence of any poles. It will coſt very little expence to fill a whole acre in that manner, making an interval of a foot or eighteen inches, in order that the ſun may have acceſs to all of them equally.

CHAP. XI.

MANURES.

THE manures commonly uſed in Scotland are dung, lime, ſhell-marl, clay-marl, and ſtone-marl. Many other ſubſtances are uſed; ſhavings

ſhavings of horn, for example, refuſe of malt, and even old rags: but as the quantity that can be procured is inconſiderable, and as their application is ſimple, I ſhall conſume no time upon them.

Dung is the chief of all manures; becauſe a quantity of it may be collected in every farm, and becauſe it makes the quickeſt return. A field ſufficiently dunged, will produce good crops four or five years.

Dung of animals that chew the cud, being more thoroughly putrefied than that of others, is fit to be mixed with the ſoil without needing to be collected into a dunghill. A horſe does not chew the cud; and in horſe-dung may be perceived ſtraw or ryegraſs broken into ſmall parts, but not diſſolved: it is proper therefore that the putrefaction be completed in a dunghill. It ought to be mixed there with cool materials: ſo hot it is, that in a dunghill by itſelf, it ſinges and burns inſtead of putrefying. The difference between the dung of a horſe and of a horned animal, is viſible in a paſture-field: the graſs round the former is withered; round the latter, it is ranker and more verdant than in the reſt of the field. A mixture of dry and moiſt ſtuff, ought to be ſtudied: the former attracting moiſture from the latter, they become equally moiſt.

To

To prevent ſap from running out of a dunghill, its ſituation ſhould be a little below the ſurface; and to prevent rain from running into it, it ſhould be ſurrounded with a ring of ſod. If the ſoil on which the dunghill ſtands be porous, let it be paved, to prevent the ſap from ſinking into the ground. If moiſture happen to ſuperabound, it may be led off by a ſmall gutter to impregnate a quantity of rich mould laid down to receve it, which will make that mould equal to good dung.

Straw ſhould be prepared for the dunghill, by being laid under cattle, and ſufficiently moiſtned. When laid dry in a dunghill, it keeps it open, admits too much air, and prevents putrefaction.

Dung from the ſtable ought to be carefully ſpread on the dunghill, and mixed with the former dung. When left in heaps upon the dunghill, fermentation and putrefaction go on unequally.

Complete putrefaction is of importance with regard to the feed of weeds that are in the dunghill: if they remain ſound, they are carried out with the dung, and infeſt the ground. Complete putrefaction is of ſtill greater importance by pulveriſing the dung; in which condition it mixes intimately with the ſoil, and operates the moſt powerfully. In land intended for barley, undigeſted dung has an unhappy effect: it keeps the ground open, admits drought, and prevents the ſeed from ſpringing. On the other hand, when thoroughly

thoroughly rotted, it mixes with the ſoil, and enables it to retain moiſture. It follows, that the propereſt time for dunging a field, is in its higheſt pulveration; at which time the earth mixes intimately with the dung. Immediately before ſetting cabbage, ſowing turnip, or wheat, is a good time. Dung divides and ſpreads the moſt accurately when moiſt. Its intimate mixture with the ſoil is of ſuch importance, that hands ſhould be employed to divide and ſpread any lumps that may be in it. Tho' dung is the chief manure in Scotland, the generality of our farmers ſeem not to give due attention to it. They are not only negligent in collecting materials, but apply it green without being putrified. It may be juſtly ſaid that the half at leaſt of its benefit is loſt; or, in other words, that the effect would be double were it well prepared and mixed intimately with the ſoil.

Dung ſhould be ſpread, and plowed into the ground, without delay. When a heap lies two or three weeks, ſome of the moiſture ſinks into the ground, which will produce tufts of corn more vigorous than in the reſt of the field. There cannot be a worſe practice than to lead out dung before winter, leaving it expoſed to froſt and ſnow. The whole ſprit of the dung is extracted by rain, and carried off with it. The dung diveſted of its ſap becomes dry in ſpring, and incapable of being mixed with the mould. It is turned over whole by the plough, and buried in the furrow.

I approve not of plowing down buck-wheat, red clover, or any other crop, for manure. The best way of converting a crop into manure is, to pass it through the body of an animal. The dung and urine, not to mention the profit of feeding, will enrich the ground more than to plow down the crop.

As dung is an article of the utmost importance in husbandry, one should imagine, that the collecting it would be a capital article with an industrious farmer. Yet an ingenious writer, observing, that the Jamaicans are in this particular much more industrious than the British, ascribes the difference to the difficulty of procuring dung in Jamaica. "In England, where the long "winter enables a farmer to raise what quantity "he pleases, it is not collected with any degree of industry. But in Jamaica, where "there is no winter, and where the heat of the "sun is a great obstruction, the farmer must be "indefatigable, or he will never raise any dung." Cool interest is not alone a sufficient motive with the indolent, to be active. As dung is of great importance in husbandry, a farmer cannot be too assiduous, in collecting animal and vegetable substances that will rot. One article of that kind there is, to collect which there is a double motive, and yet is neglected almost every where.

A farm full of weeds is a nuiſance to the neighbourhood: it poiſons the fields around; and the poſſeſſor ought to be diſgraced as a peſt to ſociety. Now the cutting down every weed before the ſeed is formed, anſwers two excellent purpoſes. Firſt, it encourages good crops by keeping the ground clean. Next, theſe weeds mixed with other materials in a dunghill, will add conſiderably to the quantity of dung *.

In erecting a large dunghill, a cart and horſes are commonly employed to lay the materials upon it; and in a dunghill of a ſmaller ſize, a hand barrow is commonly employed. This practice I cannot approve; for where a dunghill is much trodden upon, the air is excluded from the parts that are the moſt compreſſed, which prevents putrefaction. A dunghill compoſed, as it ought to be, of half digeſted materials, may be raiſed as high as can be done with the reach of the hand: its own weight will compreſs it ſufficiently for putrefaction. And to prevent evaporation, it may be finiſhed with a covering of fine earth and barley ſown on it.

Next of lime, which is a profitable manure, and greatly profitable when it can be got in plenty within a moderate diſtance. Philoſophers differ widely about its nature, and the cauſe of its

* At the roots of hedges in England fronting the high-road, the weeds grow in quantity ſufficient, if collected, to make many thouſand cart-loads of dung yearly.

its effects; and they talk so loosely, as to convince a plain farmer that the matter is very little understood. But practice is our present theme; and the benefit of lime is so visible, that the use of it has become general, where the price and carriage are moderate.

However people may differ in other particulars, all agree, that the operation of lime depends on its intimate mixture with the soil; and therefore that the proper time of applying it, is when it is perfectly powdered and the soil at the same time in the highest degree of pulveration. This opinion appears to have a solid foundation. Lime of itself is absolutely barren; and yet it enriches a barren soil. Neither of the two produces any good effect without the other: therefore the effect must depend on the mixture; and consequently the more intimately they are mixed the effect must be the greater *.

Hence it follows, that lime ought always to be slaked with a proper quantity of water, because by that means it is reduced the most effectually into powder. Lime left to be slaked by a moist air, or accidental rain, is seldom or never thoroughly reduced into powder; and therefore can never be intimately mixed with the soil. Sometimes

* Mr Buchanan of Achleshy, in Perthshire, prepared a quantity of lime for manuring a moor plowed before winter. Lime was immediately spread upon a part of it; but the work was stopped by bad weather. The remaining

times an opportunity offers to bring home ſhell-lime before the ground is ready for it; and it is commonly thrown into a heap without cover, truſting to rain for ſlaking. The proper way is, to lay the ſhell-lime in different heaps on the ground where it is to be ſpread, to reduce theſe heaps into powder by ſlaking with water, and to cover the ſlaked lime with ſod ſo as to defend it from rain. One however would avoid as much as poſſible the bringing home lime before the ground is ready for it. Where allowed to lie long in a heap, there are two bad conſequences: firſt, lime attracts moiſture, even though well covered,

maining lime was indeed covered with ſod, but ſo ſlightly as not to throw off the rain. When laid on the reſt of the field in March, it was ſo clotted as to be but imperfectly mixed with the ſoil. The crop of oats on the part firſt limed was good; that on the part laſt limed was good for nothing. The ſubſequent crops however on this part proved tolerable, the lime, by repeated plowings being better mixed with the ſoil.——In ſeveral parts of Scotland are found limeſtone of two different ſorts. The operation of the one is quick, when ſpread upon a field after calcination; but its prolific effects are ſoon over. The other operates more ſlowly; but its prolific effects continue longer. The former upon being ſlaked falls readily into a very fine powder; the latter falls more ſlowly into a powder that is not ſo fine. This accounts for the difference. The fine powder mixes more intimately with the ſoil, and more quickly, than the coarſe powder. For the ſame reaſon, the fine powder makes the beſt cement, by mixing eaſily and perfectly with the ſand and water.

vered, and runs into clots, which prevents an intimate mixture; and, next, we learn from Dr Black, that burnt limeſtone, whether in ſhells or in powder, returns gradually into its original ſtate of limeſtone; and upon that account alſo, is leſs capable of being mixt with the ſoil. And this is verified by a fact, that after lying long, it is ſo hard bound together as to require a pick-axe for breaking it down. Therefore, make it an indiſpenſable rule not to manure wet ground with lime: it will run into clots, and never mix intimately with the earth.

For the ſame reaſon, it is a bad practice, tho' common, to let ſpread lime lie on the ſurface all winter. The bad effects above-mentioned take place here in part: and there is another; that rain waſhes the lime down to the furrows, and in a hanging field carries the whole away.

As the particles of powdered lime are both ſmall and heavy, they quickly ſink to the bottom of the furrow, if care be not taken to prevent it. In that view, it is a rule, that lime be ſpread and mixed with the ſoil, immediately before ſowing, or along with the ſeed. In this manner of application, there being no occaſion to move it till the ground be ſtirred for a new crop, it has time to incorporate with the ſoil, and does not readily ſeparate from it. Thus, if turnip-ſeed is to be ſowed broadcaſt, the lime ought to be laid on immediately before ſowing, and harrowed in with the

the ſeed. If a crop of drilled turnip or cabbage be intended, the lime ought to be ſpread immediately before forming the drills. With reſpect to wheat, the lime ought to be ſpread immediately before ſeed-furrowing. If ſpread more early, before the ground is ſufficiently broken, it ſinks to the bottom. If a light ſoil be prepared for barley, the lime ought to be ſpread after ſeed-furrowing, and harrowed in with the ſeed. In a ſtrong ſoil, it ſinks not ſo readily to the bottom; and therefore, before ſowing the barley, the lime ought to be mixed with the ſoil by a brake. Where moor is ſummer-fallowed for a crop of oats next year, the lime ought to be laid on immediately before the laſt plowing, and braked in as before. It has ſufficient time to incorporate with the ſoil before the land is ſtirred again.

The quantity to be laid on, depends on the nature of the ſoil. Upon a ſtrong ſoil, ſeventy or eighty bolls of ſhells are not more than ſufficient, reckoning four ſmall firlots to the boll, termed wheat-meaſure; nor will it be an over-doze to lay on a hundred bolls. Between fifty and ſixty may ſuffice upon medium ſoils; and upon the thin or gravelly, between thirty and forty. It is not ſafe to lay a much greater quantity on ſuch ſoils.

It is common to lime a paſture-field immediately before plowing. This is an unſafe practice: it is thrown to the bottom of the furrow, from which

which it is never fully gathered up. The proper time for liming a paſture-field, intended to be taken up for corn, is a year at leaſt, or two, before plowing. It is waſhed in by rain among the roots of the plants ; and has time to incorporate with the ſoil.

With regard to the expence of carriage, to have the lime-kiln ſo near as to go twice a-day, is a great ſaving. But if there can be but one carriage in a day, there is little difference as to expence, whether the diſtance be ſeven or eleven miles. A little more food to the cattle makes all equal.

Limeſtone beat ſmall makes an excellent manure, and ſupplies the want of powdered lime, where there is no feuel to burn the limeſtone. Limeſtone beat ſmall has not hitherto been much uſed as a manure ; and the proportion between it and powdered lime has not been aſcertained. What follows may give ſome light. Three pounds of raw lime is by burning reduced to two pounds of ſhell-lime. Yet nothing is expelled by the fire but the air that was in the limeſtone : the calcarious earth remains entire. *Ergo*, two pounds of ſhell-lime contain as much calcarious earth as three pounds of raw limeſtone. Shell-lime of the beſt quality, when ſlaked with water, will meaſure out to thrice the quantity. But as limeſtone loſes none of its bulk by being burnt into ſhells, it follows, that three buſhels of raw limeſtone

limeſtone contain as much calcarious earth as ſix buſhels of powdered lime. And conſequently, if powdered lime poſſeſs not ſome virtue above raw limeſtone, three buſhels of the latter beat ſmall ſhould equal as a manure ſix buſhels of the former.

Shell-marl, as a manure, is managed in every reſpect like powdered lime, with this only difference, that a fifth or a fourth part more in meaſure ought to be given. The reaſon is, that ſhell-marl is leſs weighty than lime, and that a boll of it contains leſs calcarious earth, which is the fructifying part of both.

I ſhall conclude with clay and ſtone marls, which, with reſpect to huſbandry, are the ſame, though in appearance different. The manures hitherto mentioned are reſtoratives only: they recruit land when worn out by cropping, and enable it to bear more crops. The marl now under conſideration is not only a reſtorative, but has an effect ſtill more deſirable, that of altering the nature of the ſoil, and improving its texture, ſo as to convert it from light to heavy, and from weak to ſtrong. I know nothing comparable to it in that reſpect, but the poaching light land by cattle fed with turnip, mentioned m a foregoing chapter. It has another effect, in appearance oppoſite; which is, to looſen a clay ſoil, and to make it more free.

The goodneſs of this marl depends on the

quantity of calcarious earth in it; which I have known to amount to a half or more. It is too expensive if the quantity be less than a third or a fourth part. Good marl is the most substantial of all manures; because it improves the weakest ground to equal the best borough-acres. One instance I know, of two ridges marled a hundred and twenty years ago, that at this day make a figure both in grass and corn far above the rest of the field. The low part of Berwickshire, termed *the Merse*, abounds every where with this marl; and in no other part of Scotland is it in such plenty.

As none of the manures I am acquainted with make any distinction between weeds and corn, the land ought to be cleared of weeds before marling; and it ought to be smoothed with the brake and harrow, in order that the marl may be equally spread. Marl is a fossil on which no vegetable will grow: its efficacy depends, like that of lime, on its pulveration, and intimate mixture with the soil. Toward the former, alternate drought and moisture contribute greatly, as also frost. Therefore after being evenly spread, it ought to lie on the surface all winter. In the month of October, it may be roused with a brake, which will bring to the surface, and expose to air and frost, all the hard parts, and mix with the soil all that is powdered. In that respect it differs widely from dung and lime, which ought

to

to be plowed into the ground without delay. Oats is a hardy grain, which anſwer for being the firſt crop after marling, better than any other; and it will ſucceed though the marl be not thoroughly mixed with the ſoil. In that caſe, the marl ought to be plowed in with an ebb furrow immediately before ſowing, and braked thoroughly. It is tickliſh to make wheat the firſt crop: if ſown before winter, froſt ſwells the marl, and is apt to throw the ſeed out of the ground; if ſown in ſpring, it will ſuffer more than oats by want of due mixture.

Summer is the proper ſeaſon for marling; becauſe in that ſeaſon the marl, being dry, is not only lighter, but is eaſily reduced to powder. Froſt however is not improper for marling, eſpecially as in froſt there is little opportunity for any other work.

Marl is a heavy body, and ſinks to the bottom of the furrow, if indiſcreetly plowed. Therefore the firſt crop ſhould always have an ebb furrow. During the growing of that crop, the marl has time to incorporate with the ſoil, and to become a part of it; after which it does not readily ſeparate.

Stone marl is ſo hard by a conſiderable mixture of ſand, that it will continue without diſſolving for years. In that caſe, the expence of breaking the larger lumps with hammers would not be loſt: ſuch lumps have no effect to promote vegetation;

getation; and they are beſide an obſtruction to plowing and harrowing.

About twenty years ago, many Merſe tenants applied their whole ſtrength to marling, with very great ſucceſs: and yet of late they ſeem more intent on liming; which may appear ſingular in a country where huſbandry goes on with alacrity. But leaſes in that country are commonly limited to nineteen years, which, it is thought, affords too little time for drawing all the profit from this expenſive manure, that the tenant is intitled to. To marl to perfection, requires four hundred cart-load on an acre, as much as can be drawn by two ſturdy horſes; which, at a moderate computation, coſts about L. 4 *per* acre. But a field can be ſufficiently limed from the diſtance of fifteen Engliſh miles, for little above the half of that ſum; with this additional convenience, that a farmer can hire carts for liming, inſtead of being cofined to his own horſes, as he muſt be in marling; by which means, liming can be carried on with much more expedition than marling. Notwithſtanding theſe differences, it is eaſy to evince, that even on a leaſe of nineteen years, marl will afford a greater profit than lime. Limed ground cannot bear without injury above three or four crops; after which it ought to be laid down in graſs. A field well marled, will produce rich crops of corn, in the ſimpleſt manner of culture, as long as the leaſe endures. Now,

though

though graſs is more profitable than corn in a poor ſoil, the profit bears no proportion in a rich ſoil, which produces excellent crops, with no greater labour nor expence than is neceſſarily beſtowed on poor ſoil.

CHAP. XII.

FENCES.

IN Scotland, fences of ſeveral kinds are uſed. Stone-walls and thorn-hedges are the beſt and the moſt common.

The height of a dry-ſtone wall is directed by the uſe it is intended for. If intended for ſheep, it cannot be under ſix feet high, every rood of which, being thirty-ſix ſquare ells, will at a medium coſt twenty ſhillings. A dry-ſtone wall for horſes or horned cattle cannot be under five feet high. The expence of incloſing in this manner is conſiderable. A ſquare field of ten acres incloſed with a wall ſix feet high, will coſt L. 50, 15s.; and L. 42 : 5 : 10, if the wall be five feet high. And it will require two and one half *per cent.* annually to keep them in order. To build with lime, as many do, inflames the coſt; and yet upon the whole is a ſaving where lime is at hand.

To reduce both the expence of building dry-ſtone walls, and of upholding them, I warmly recommend

commend the following mode. Raiſe the wall to the height only of two feet and a half, and cope it with ſod in the following manner. Firſt, lay on the wall with the graſſy ſide under, ſod cut with the ſpade four or five inches deep, and of a length to equal the thickneſs of the wall. Next, cover this ſod with looſe earth rounded like a ridge. Third, prepare thin ſod, caſt with the paring ſpade, ſo long as to extend beyond the thickneſs of the wall two inches on each ſide. With theſe cover the looſe earth, keeping the graſſy ſide above: place them ſo much on edge as that each ſod ſhall cover part of another, leaving only two inches without cover. Fourth, when twenty or thirty yards are thus finiſhed, let the ſod be beat with mells by two men, one on each ſide of the wall, ſtriking both at the ſame time. By this operation, the ſod becomes a compact body that keeps in the moiſture, and encourages the graſs to grow. Laſtly, cut off the ragged ends of the ſod on each ſide of the wall, to make the covering neat and regular. The month of October is the proper ſeaſon for this operation, becauſe the ſun and wind, during ſummer, dry the ſod and hinder the graſs from vegetating. Moiſt ſoil affords the beſt ſod. Wet ſoil is commonly too fat for binding; and at any rate, the watery plants it produces will not thrive in a dry ſituation. Dry ſoil, on the other hand, being commonly ill bound with roots, ſhakes to

pieces

pieces in handling. The ordinary way of coping with ſod, which is to lay them flat and ſingle, looks as if intended to dry the ſod and to kill the graſs ; not to mention, that the ſod is liable to be blown off the wall by every high wind.

Having finiſhed the wall, make a ditch on each ſide, beginning a foot from the root of the wall, and ſloping outward to the depth of three feet, or at leaſt two and a half. The ditch ſhould be equally ſloped on the other ſide, ſo as to be four feet wide.

A rood of this fence, including every article, may be done for three ſhillings or thereabout ; and a field of ten acres may be thus incloſed for about L. 30. If the ditch be made three feet deep, the fence will be above ſix feet high ; and above L. 20 will be ſaved of what a dry-ſtone wall ſix feet high will coſt. Nor is the ſaving of this ſum the moſt conſiderable article. A fence of which the parts are ſo well connected, will ſtand many years with little or no reparation. That this is far from being the caſe of a dry-ſtone wall ſix feet or even five feet high, all the world know. Though a deer-park of great extent, ſuch as we ſee many in England, is no favourite of mine, I have no objection to one of forty or fifty acres ; which may be incloſed at a very ſmall expence. After ſurrounding it with the wall here deſcribed, plant within laburnums cloſe to the wall. Lope off their heads to make the branches

branches extend laterally and interweave in form of a hedge. The wall will prevent the deer from breaking through; and if the hedge be trained eight feet high, they will not attempt to leap over. I prefer the laburnum, becauſe no beaſt will feed on it, except a hare, and that only when young and the buſh tender. Therefore no extraordinary care is neceſſary, except to preſerve them from the hare four or five years. A row of alders may be planted in front of the laburnums, which a hare will not touch nor any other beaſt.

Next of thorn-hedges. The advantage of the white thorn for a fence above every other plant, is well underſtood. It is a quick grower when planted in proper ſoil, ſhooting up ſix or ſeven feet in a ſeaſon. Though tender when young and hurt by weeds, it turns ſtrong, and may be cut into any ſhape. Even when old, it is more diſpoſed than other trees to lateral ſhoots. And, laſtly, its prickles make it the moſt proper of all for a fence.

The method of ſowing the ſeed in beds, is ſet forth in the chapter of Foreſt-Trees. After the plants have ſtood a full year in the ſeed-bed, tranſplant the largeſt into a nurſery, which will leave ſufficient room for the remainder, to ſtand where they are another year. In the nurſery, they ought to ſtand at the diſtance of ſeven, eight, or nine inches from each other; and there they may remain till fit to be planted in a hedge, which is no ſooner than at the age of five. Room in a nurſery

nurſery is of great importance: when ſtraitened for room, the plants ſhoot up faſt, are weakly, and unfit to bear the hardſhips of an open field. The diſtance ought to be proportioned to the ſoil; the greateſt in a rich ſoil, becauſe they grow faſt; the leaſt in a poor ſoil, where they grow ſlow. The beſt ſoil for a nurſery is between rich and poor. In the latter, the plants are dwarfiſh: in the former, being luxuriant and tender, they are apt to be hurt during the ſeverity of weather: and theſe imperfections are incapable of any remedy. An eſſential requiſite in a nurſery is free ventilation. How common is it to find nurſeries in hollow-ſheltered places, ſurrounded with walls and high plantations, more fit for pine-apples than for barren trees! The plants thruſt out long ſhoots, but feeble and tender: when expoſed to a cold ſituation, they decay, and ſometimes die. But there is a reaſon for every thing: the nurſeryman's view is to make profit by ſaving ground, and by impoſing on the purchaſer tall plants, for which he pretends to demand double price. It is ſo difficult to purchaſe wholeſome and well-nurſed plants, that every gentleman-farmer ought to raiſe plants for himſelf.

As thorns will grow pleaſantly from roots, I have long practiſed a frugal and expeditious method, of raiſing them from the wounded roots that muſt be cut off when thorns are to be ſet in a hedge. Theſe roots cut into ſmall parts, and put in a bed of fresh earth, will produce plants the next spring, no less vigorous than what are produced from seed. And thus a perpetual succession of plants may be obtained without any more seed.

It ought to be a rule, never to admit into a hedge plants under five years old: they deserve all the additional sum that can be demanded for them. Young and feeble plants in a hedge, are of slow growth; and beside loss of time, the paling, necessary to secure them from cattle, must be renewed more than once before they become a fence.

A thorn-hedge may be planted in every month of winter and spring, unless it be frost. But I have always observed, that thorns planted in October are more healthy, push more vigorously, and fewer decay, than at any other time. In preparing the thorns for planting, the roots ought to be left as entire as possible, and nothing cut away but the ragged parts.

As a thorn-hedge suffers greatly by weeds, the ground where they are to be planted ought to be made perfectly clean. The common method of planting is, to leave eight or nine inches along a side of the intended ditch, termed a scarsement; and behind the scarsement to lay the surface-soil of the intended ditch, cut into square sods two or three inches deep, its grassy surface under. Upon that sod, whether clean or dirty, the

the thorns are laid, and the earth of the ditch above them. The graſs in the ſcarſement, with what weeds are in the moved earth, ſoon grow up, and require double diligence to prevent the young thorns from being choked. The following method, which is creeping into practice, deſerves all the additional trouble it requires. Leaving a ſcarſement as above of ten inches, and alſo a border for the thorns, broad or narrow according to their ſize, lay behind the border all the ſurface of the intended ditch, champed ſmall with the ſpade, and upon it lay the mouldery earth that fell from the ſpade in cutting the ſaid ſurface. Cover the ſcarſement and border with the under earth, three inches thick as leaſt; laying a little more on the border to raiſe it higher than the ſcarſement, in order to give room for weeding. After the thorns are prepared, by ſmoothing their ragged roots with a knife, and lopping off their heads in order to make them grow buſhy, they are laid fronting the ditch, with their roots on the border, the head a little higher than the root. Special care muſt be taken to ſpread the roots among the ſurface-earth taken out of the ditch, and to cover them with the mouldery earth that lay immediately below. This article is of importance, becauſe the mouldery earth is the fineſt of all. Cover the ſtems of the thorns with the next *ſtratum* of the ditch, leaving always an inch at the top free. It is no matter how

how poor this *ſtratum* be, as the plants draw no nouriſhment from it. Go on to finiſh the ditch, preſſing down carefully every row of earth thrown up behind the hedge, which makes a ſolid mound impervious to rain. It is a ſafeguard to the young hedge to raiſe this mound as perpendicular as poſſible; and for that end it may be proper in looſe ſoil, when the mound is raiſed a foot or ſo, to bind it with a row of the tough ſod, which will ſupport the earth above, till it become ſolid by lying.

This is ſufficient in rich ſoil; but in poor ſoil greater care is neceſſary. Behind the line of the ditch, the ground intended for the ſcarſement and border ſhould be ſummer-fallowed, manured, and cleared of all graſs-roots; and this culture will make up for the inferiority of ſoil. In very poor ſoil it is vain to think of planting a thorn-hedge; for it will give no ſatisfaction, becauſe it never will become a fence. In ſuch ground there is a neceſſity for a ſtone fence.

The only reaſon that can be given for laying thorns as above deſcribed, is to give the roots ſpace to puſh in all directions, even upward into the mound of earth. There may be ſome advantage in this; but, in my apprehenſion, the diſadvantage is much greater of heaping ſo much earth on the roots, as to exclude not only the ſun and air, but even rain which runs down the ſloping bank, and has no acceſs to the roots. Inſtead

ſtead of laying the thorns fronting the ditch, would it not do better to lay them parallel to the ditch; covering the roots with three or four inches of the beſt earth, which would make a hollow between the plants and the ſloping bank? This hollow would intercept every drop of rain that falls on the bank, to ſink gradually among the roots. If this be not a better poſition for a thorn, it muſt be of a ſingular conſtitution. I venture one ſtep farther out of the common path. Why at any rate ſhould a thorn be put in the ground ſloping? This is not the practice with reſpect to any other tree; and I have heard of no experiment to perſuade me that a thorn thrives better ſloping than erect. But as in natural hiſtory experiment is always our laſt reſource, I am at preſent following out a comparative trial. I have many young hedges in the way firſt deſcribed: ſome I have laid parallel to the ditch in the way now deſcribed, and ſome I have planted erect with their tops entire. The trials are fairly made; and time will determine the beſt method. There indeed occurs one objection againſt planting the thorns erect, that the roots have no room to extend themſelves on that ſide where the ditch is. But does it not hold that when in their progreſs roots meet a ditch, they do not puſh onward, but changing their direction puſh downward at the ſide of the ditch? (See appendix article 4th.) If ſo, theſe downward roots will ſup-

port

port the ditch, and prevent it from being mouldered down by froſt.

One thing is evident without experiment, that thorns planted erect may ſooner be made a complete fence than when laid ſloping as uſual. In the latter caſe, the operator is confined to thorns that do not exceed a foot or 15 inches; but thorns five or ſix feet high may be planted erect and a hedge of ſuch thorns, well cultivated in the nurſery, will in three years arrive to greater perfection, than a hedge managed in the ordinary way will do in twice that time.

In a rich ſoil the thorns may be planted ten inches aſunder; and from that to ſix inches in inferior ſoils. To preſerve them from cattle, the ditch ought to be ſix feet wide, three feet deep, and as narrow at bottom as the breadth of the ſpade will allow. But when the thorns are ſufficiently covered, the prudent way is to delay finiſhing the ditch till the froſt be over. Spring is the proper time for completing the work, while there is ſufficient ſap in the ſoil to bind the mound of earth that is raiſed behind the thorns. If delayed till the earth be dry, it never binds: if completed the beginning of winter, froſt ſwells the earth, looſens it, and makes it crumble down.

The hedge is fenced from cattle on the one ſide by the ditch; but it is neceſſary that it be fenced on both ſides. The ordinary method of a paling is no ſufficient fence againſt cattle: the moſt gentle

gentle make it a rubbing-poſt; and the vitious break it down wantonly with their horns. The only effectual remedy is expenſive; but better no fence than one that is imperfect. The remedy is, two ditches and two hedges, with a high mound of earth between them. Survey Scotland, and you will not find any fences otherwiſe conſtructed without many breaches; the reparing of which gives much trouble and little ſatisfaction. If this remedy however be not palatable, the paling ought at leaſt to be of the ſtrongeſt kind. Many different kinds have been put in practice, that are extremely frail. I recommend the following, as the beſt I am acquainted with. Drive into the ground ſtrong ſtakes three feet and a half long, with intervals from eight to twelve inches, according to the ſize of the cattle that are to be encloſed; and all preciſely of the ſame height. Prepare plates of wood ſawed out of logs, every plate three inches broad and half an inch thick. Fix them on the head of the ſtakes, with a nail driven down into each ſtake. The ſtakes will be united ſo firmly, that one cannot be moved without the whole; and will be proof accordingly againſt the rubbing of cattle. But, after all, it is no fence againſt vitious cattle. The only proper place for it is the ſide of a high-road, or to fence a plantation of trees. It will indeed be a ſufficient fence againſt ſheep, and endure till the hedge itſelf be a fence. A fence thus completed, ed, including thorns, ditching, wood, nails, *&c.* will not much exceed two ſhillings every ſix yards.

We are now arrived at the moſt important article of all, that of training up a thorn-hedge after it is planted. The ordinary method is, to cut off the top and ſhorten the lateral branches, in order to make it thick and buſhy. To the ſame end, the young thorns, after ſtanding ſix or ſeven years, are ſometimes cut over within two or three inches of the ground, which multiplies the ſtems, and makes the hedge ſtill thicker. This form of a hedge catches the eye: by its thickneſs it is formidable to cattle; but its weakneſs is diſcovered when bare of leaves, and cattle break through every where without obſtruction.

I have the experience of three hedges trained twelve years as follows. The firſt has been annually pruned top and ſides. The ſides of the ſecond have been pruned, but the top left entire. The third was allowed to grow without any pruning. The firſt is at preſent about four feet broad and thick from top to bottom; but weak in its ſtems, and unable to reſiſt any horned beaſt. The ſecond is ſtrong in its ſtems, and cloſe from top to bottom. The third is alſo ſtrong in its ſtems, but for two feet up bare of lateral branches, which have been deſtroyed by the overſhadowing of thoſe above, depriving them both of rain and air.

air. That the ſecond is the beſt method, is aſcertained by experience. And that it ought to be ſo, will be evident if we can truſt to analogy: in the natural growth of a tree, its trunk is proportioned to its height: lop off the head, it ſpreads laterally and becomes a buſh, without riſing in height, or ſwelling in the trunk. The lime-tree is the only exception I know: the white thorn I am certain is not an exception. Hence the following method of training up a hedge, which is to allow the thorns to grow without applying a knife to their tops, till their ſtems be five or ſix inches in circumference. In good ſoil with careful weeding, they will be of that bulk in ten or twelve years, and be fifteen feet high or upward. The lateral branches only muſt be attended to. Thoſe next the ground muſt be pruned within two feet of the ſtem; thoſe above muſt be made ſhorter and ſhorter, in proportion to their diſtance from the ground; and at five feet high they muſt be cut cloſe to the ſtem, leaving all above full freedom of growth. By this dreſſing, the hedge takes on the appearance of a very ſteep roof; and it ought to be kept in that form by pruning. This form gives free acceſs to rain, ſun, and air: every twig has its ſhare, and the whole is preſerved in vigour. When the ſtems have arrived at their proper bulk, cut them over at five feet from the ground, where the lateral branches end. This anſwers two excellent purpoſes. The firſt is, to ſtrengthen the hedge, the ſap that formerly aſcended to the top being now diſtributed among the branches. The next is, that a tall hedge ſtagnates the air, and poiſons both corn and graſs near it. A hedge trained up in this manner, is impenetrable even by a bull: he may preſs in the lateral branches; but the ſtems ſtand firm. For an inſtant proof that this method will anſwer, obſerve the thorns that from ſpace to ſpace are allowed to grow up above their fellows in form of a hedge-row. Theſe thorns, though growing in the middle of a buſhy hedge, have ſtems far larger than the reſt. Beſide the ſtrength of ſuch a hedge, it is leſs expenſive than a hedge reared in the ordinary way: the weeds are ſooner choaked; and it requires much leſs pruning. If the ground have been prepared for the plants by cleaning it perfectly, the hedge may be ſafely left to itſelf for four or five years; unleſs it be to ſhorten any luxuriant branch that grows much faſter than the reſt. When ſo much labour is ſaved, one will the leſs grudge the price of the beſt thorns that can be procured. Good thorns are indeed more eſſential in this manner of training a hedge than in any other: they ought all to be of an equal ſize, and equally vigorous, that they may not overtop one another. The intermixing of ſtrong and weak plants is of leſs importance, where the heads are cut off and all made to grow equally.

Plaſhing

Plashing an old hedge, an ordinary practice in England, makes indeed a good interim fence; but at the long-run is destructive to the plants: and accordingly, there is scarce to be met with a complete good hedge where plashing has been long practised. A cat is said among the vulgar to have nine lives: is it their opinion, that a thorn, like a cat, may be cut and slashed at will without suffering by it? A thorn is a tree of long life. If instead of being massacred by plashing, it were raised and dressed in the way here described, it would continue a firm hedge perhaps five hundred years. This merits attention.

A hedge ought never to be planted on the top of the mound of earth thrown up from the ditch. It has indeed the advantage of an awful situation; but being planted in bad soil, and destitute of moisture, it cannot thrive: it is at best dwarfish, and frequently decays and dies.

To plant trees in the line of the hedge, or within a few feet of it, ought to be absolutely prohibited, as a pernicious practice. It is amazing, that people should fall into this error, when they ought to know, that there never was a good thorn-hedge with trees in it. And how should it be otherwise? an oak, a beech, an elm, grows faster than a thorn: when suffered to grow in the midst of a thorn-hedge, it spreads its roots every where, and robs the thorns of their nourishment. Nor is this all: the tree overshadowing the thorns

thorns keeps the ſun and air from them. At the ſame time, no tree takes worſe with being overſhadowed than a thorn *.

It is ſcarce neceſſary to mention gaps in a hedge; becauſe they will ſeldom happen where a hedge is trained as above recommended. But in the ordinary method of training, gaps are frequent, partly by the failure of plants, and partly by the treſpaſſing of cattle. The ordinary method of making up gaps is, to plant ſweet brier where the gap is ſmall, and a crab where it is large. This method I cannot approve, for an obvious reaſon: a hedge ought never to be compoſed of plants that grow unequally: thoſe that grow faſt overtop and hurt the ſlow growers; and with reſpect in particular to a crab and ſweet brier, neither of them thrive under ſhade. It is a better method to remove all the withered earth in the gap, and to ſubſtitute freſh ſappy mould mixed with ſome lime or dung. Plant upon it a vigorous thorn of equal height with the hedge, which in its growth will equal the thorns it is mixed with. In that view, there ſhould be a nurſery

* In England there is ſcarce a thorn-fence to be ſeen without a hedge-row of trees; and theſe hedge-rows have been the deſtruction of fences. The thorns, as far as the roots or branches of the trees extend, muſt decay. It is in vain to put young thorns in the gap, for they will not take root. To fill up gaps, plaſhing became neceſſary. Plaſhing for a time makes a ſtrong fence; and by that means became general.

nurſery of thorns of all ſizes, even to five feet high, ready to fill up gaps. The beſt ſeaſon for this operation is the month of October. I ſhould have added, that a gap filled with ſweet brier, or a crab lower than the hedge, invites cattle to break through and trample the young plants under foot; to prevent which a paling on both ſides is not ſufficient, unleſs it be raiſed as high as the hedge *.

With reſpect to a field that is too poor for thorns, what ſhall be the fence if there be no ſtones in the neighbourhood? In ſuch fields, a hedge of whins is the only reſourſe. Whin-hedges ſtand commonly on the top of a dry earth-dyke, in which ſituation they ſeldom thrive well. I like what follows better. Two parallel ditches three feet wide and two deep, border a ſpace of twelve feet. Within this ſpace, raiſe a bank at the ſide of each ditch with the earth that comes out of it, leaving an interval between the two banks. Sow the banks with whin ſeed, and plant a row of trees in the interval. When the whins are pretty well grown, the hedge on one of the banks may be cut down; then the other as

* In ſeveral parts of England, a vicious practice has crept in of cutting off the loweſt lateral branches, leaving a buſhy top, but the ſtem bare. This I preſume muſt be in imitation of a tree, for it cannot otherwiſe be accounted for; and yet the general practice is to avoid the form of a tree as much as poſſible.

as ſoon as the firſt becomes a fence; and ſo on alternately. While the whins are young, they will not be diſturbed by the cattle, if paſſages be left to go out and in. Theſe paſſages may be cloſed up, when the hedge is ſufficiently ſtrong to be a fence. A whin-hedge thus managed, will laſt many years, even in ſtrong froſt unleſs very ſevere. There are many whin hedges in the ſhire of Kincardine, not ſo ſkilfully managed; and yet the poſſeſſors appear not to be afraid of froſt. Such fences ought to be made extremely welcome in the ſandy grounds of the ſhire of Murray, where there is ſcarce a ſtone to be found. The few earth fences that are there raiſed, compoſed moſtly of ſand, very ſoon crumble down.

Nothing hitherto has been ſaid about an encloſure for ſheep. To carry a farm to its perfection, every encloſure ought to be made ſo as to keep in ſheep, though not chiefly intended for them; for ſheep ought to be mixed with other cattle in every paſture-field: they eat what others leave, and others eat what they leave. But farmers generally take a humbler flight, and are ſatisfied with one or two encloſures for ſheep. In that view I premiſe, that of all animals ſheep naturally take the wideſt range in feeding; for which reaſon, an encloſure for them ought never to be under fifty or ſixty acres. Where they have ſo much room, any fence will keep them in: where confined

confined to five, ſix, or even ten acres, the moſt awful fence is ſcarce ſufficient. I talk of the common run of ſheep; not of the large Lincolnſhire breed, habituated to encloſures, which have no genius for leaping.

Where ſtones are to be had, the cheapeſt and moſt effectual is the ſtone fence above deſcribed. If a farm cannot afford an encloſure ſo large as that now mentioned, the deepneſs of the ditch ought to be in a reciprocal proportion to the ſize of the encloſure: in a ſmall encloſure, the fence ought at leaſt to be ſix feet high. If the farmer be reduced to a quickſet hedge by want of ſtones, the ſcarſement muſt be eighteen inches broad, in order to receive a paling between the root of the hedge and lip of the ditch; inclining to the ditch in form of a ſtocade, which will make it the more awful. This will fit the encloſure immediately for ſheep, without waiting the ſlow growth of the hedge. When the paling fails, the ditch muſt be lined up with ſtones to the height of the hedge, to prevent the ſheep from making a road up the ſide of the ditch to come at the thorns; which would ruin all, for no food is ſo palatable to them as the leaves of thorn. Where ſtones are ſcarce, two feet of ſtones may do, raiſing the lining to the height of the thorns with ſod.

One thing is never to be omitted in a large encloſure. Sheep require a ſhelter againſt heat, no leſs than againſt cold. The ſtone fence recommended,

mended, affords little ſhelter againſt either. Therefore, a ſhelter ought to be made ſomewhere in the drieſt part of the field. Plant ten or twelve rows of Scotch fir or ſpruce fir in a curve, ſemicircular, or eliptical, the concave fronting the ſouth, and ſufficiently capacious for beds to all the ſheep. Surround this plantation with a ſtone fence of the ſame form with what encloſes the field. The weakeſt trees ought to be weeded out from time to time, which will give the trees that ſtand room to extend their lateral branches. And the plantation when grown up will protect the ſheep both from ſun and wind. The dung dropt round this ſheltered place, may be gathered and put in a dunghill, or ſpread upon the field.

A ſheep yields more profit by paſture than any other animal; and as its dung enriches the ſoil more, I muſt repeat again, that a provident farmer ought to have more encloſures than one: he will have the benefit of fine crops of corn by removing ſucceſſively ſheep from one encloſure to another.

With reſpect to the ſize of encloſures, ſeveral conſiderations enter. Sheep naturally take a compaſs in feeding: if reſtrained by a ſmall encloſure, it will require a ſtrong fence to keep them in: if they have ſcope in an encloſure of 30 or 40 acres, they never think of going out. An encloſure of 20 or 25 acres is ſufficient for

horſe

horſes or for horned cattle. Encloſures of a large ſize have the advantage of free ventilation, eſſential while they are in corn. There are alſo other conſiderations to be kept in view in determining the ſize of an encloſure. One is, that the ditches ought to be ſo directed as to carry off ſuperfluous moiſture. Another, that different ſoils in the ſame encloſure ought to be avoided. As heat and ſhelter only ſeem to have been formerly thought of in the Engliſh encloſures, they are generally too ſmall; by which that country has ſuffered greatly, labour loſt, land loſt, corn and graſs loſt within ten yards of the fence; and in a tickliſh ſeaſon, perhaps the whole loſt for want of ventilation. The oppreſſion of flies and ſtagnation of air, make ſummer paſture incommodious in a ſmall encloſure. It indeed gives much ſhelter in winter; but that benefit is confined to a light ſoil, for loam or clay is apt to be poached by winter-paſture.

A word more before I end, which will be on the gates of encloſures. The poſt upon which the gate is hung, though built of ſtone and lime, is apt to be ſhaken by a heavy gate, and to be torn to pieces by a careleſs driver running the axle of his cart againſt it. But inſtead of building a wall in the line of the encloſure, let it be built perpendicular to it, ſeven or eight feet long, thickeſt in the middle and tapering toward the

 ends.

ends. Such a wall or post will be proof against any force.

CHAP. XIII.

The proper size of a Farm, and the useful accommodations it ought to have.

TO confine a tenant to that quantity of land which can be managed to the best purpose with the least expence, is the surest means of obtaining an adequate rent without oppression. A farm ought never to contain a less quantity of land than sufficient for a plough; and there is no medium between that quantity, and as much as will give full employment to two. Less than sufficient for a plough, is an evident loss to the tenant, and consequently to the landlord: the servants and cattle must at times be idle for want of work; or, what is worse, they will work indolently, and make no progress. I have a striking proof of this observation. The estate of Grange in the carse of Falkirk, containing about 360 Scotch acres, paid of rent L. 450. It was possessed by no fewer than eleven tenants; not one of the farms sufficient for a plough, several of them between twenty and thirty acres. A map was made: it was clearly seen, that the estate could hold but six tenants; and it luckily happened, that there were

were ſix houſes abundantly centrical for theſe tenants, without neceſſity of new buildings. The ſaving of twenty horſes and ten ſervants, and the maintenance of five tenants and their families, was great; beſide the profit that each tenant was intitled to for his ſkill and labour. The ſix tenants that took the whole offered at once an advance of L. 194 Sterling, which was accepted, though not equal to the ſaving. And accordingly the preſent rent of L. 644 will be eaſier on the ſix tenants, than the former rent of L. 450 was on the eleven. A farm, on the other hand, that cannot be commanded by one plough, and is not ſufficient for two, is a ſtill greater loſs. The tenant, ſtruggling to make ſomething of every field, does juſtice to no field; and the farm turns poorer and poorer every year. It is well known, that moſt farms in Scotland are too large for the ability of the tenant. The reaſon is, that in an uncultivated country, the adding twenty or thirty acres more is little regarded. I could give many inſtances of a tenant beginning to thrive after being deprived of part of his farm. But it is unneceſſary to give inſtances, for they are known every where.

When one thinks of calculating what quantity of land is ſuffiicient for a ſingle plough, many circumſtances occur, that make it impracticable to determine the point with preciſion. The difference between a light and a heavy ſoil, is conſiderable;

ſiderable ; and no leſs ſo, the nearneſs or diſtance of manure. The mode of cropping is capital: where all the different plants are ſown in ſpring and reaped in autumn, more ploughs are requiſite than when crops are diſperſed through the year, according to the preſent improved mode of culture. I venture only in general to ſay, that in moſt ſoils fifty acres of corn may be commanded by a ſingle plough, provided the crops be diſtributed through the year, to afford time for managing all of them with the ſame men and cattle. But where graſs is neceſſary for keeping the ſoil in heart, a farm ought to be enlarged in proportion to the quantity of graſs required ; for there ought always to be as much land in tillage as fully to occupy a plough. If a third part in graſs be ſufficient, the farm ought to conſiſt of ſeventy-five acres; if a half be neceſſary, the farm ought to be 100 acres ; and if the ſoil be ſo mean as to require two-thirds in graſs, the extent of the farm ought to be 150 acres. If the reader be curious about further particulars, I refer him to the ſeventh chapter, in which a few examples are given of the number of acres that in different modes of cropping may be commanded by one plough. Theſe examples will pave the way to computations correſponding to other rotations.

I proceed to an intereſting article, which is to compare great and ſmall farms in point of utility.

I

I call a ſmall farm what employs but a ſingle plough ; and a ſmaller there ought not to be. A middling farm is what requires two ploughs; and whatever requires a greater number, I call a great farm. Theſe different farms I ſhall conſider with reſpect to the landlord, with reſpect to the tenant, and with reſpect to the public.

With reſpect to the landlord, there are advantages and diſadvantages that tend to balance each other. Small farms draw the greateſt number of candidates; and conſequently to produce the higheſt rents. On the other hand, ſmall farms occaſion a great expence for houſes; and in a country where building materials are coſtly, large farms may appear to be the intereſt of the landlord.

With reſpect to tenants, a farm as large as can accurately be managed, is undoubtedly the intereſt of a tenant, provided he have a fund for ſtocking the farm ſufficiently. But this is really ſaying no more, but that it is beneficial to have a large fund. The proper queſtion is, Whether with reſpect to farmers in general, it is not a convenience to have the choice of ſmall or great farms according to their ſtock? In that view, ſmall farms are undoubtedly advantageous to thoſe who want to be farmers; becauſe in Scotland, at leaſt, the number is much greater of thoſe who can ſtock a ſmall farm, than of thoſe who can go farther. It may be poſſibly objected, that there

there is an inconvenience in a ſmall farm where two horſes only are neceſſary for a plough, in reſpect that two horſes make but a ſlow progreſs in carrying corn or dung. To this objection there is a ready anſwer: two horſes in two ſingle carts will make as much expedition in carrying out the dung, or carrying in the corn of a ſmall farm, as double that number will make in a middling farm, where the dung and corn are double in quantity. I ſay further, that if two horſes be not ſufficient, the defect may be readily ſupplied by two draught oxen, which add very little to the expence of the farm. Theſe at four years of age may be purchaſed for L. 10. They will give at ſeven L. 15; and this profit, with no more work than ſufficient to give them a ſtomach, will balance their ſummer-food of green clover. Their winter food of ſtraw cannot enter into the computation, being the very beſt way of converting ſtraw into dung. Here there is a great convenience. Where a field, by drought or otherwiſe, is rendered too ſtiff for a pair, the oxen may be yoked in the plough with the horſes. In ground leſs ſtiff, the farmer has a choice of two oxen and a horſe, of two horſes alone, or of two oxen. Where plowing happens to be retarded by bad weather, two ploughs may be employed, which is a ſignal convenience. Plowing alſo and harrowing may go on at the ſame time; and the farmer has it always in his power

to

to yoke two double carts. Even in carſe-ſoil this plan will anſwer; as there is ſeldom occaſion to employ the oxen but where the ground is ſufficiently dry for them.

Wih reſpect to the public, ſmall farms are undoubtedly the moſt advantageous. The number of ſervants, it is true, muſt be in proportion to the ſize of the farm: but in a middling farm there is but one tenant; whereas in two ſmall farms of no greater extent, there are two. And the difference is ſtill greater in large farms. This is a capital circumſtance. The children of tenants are taught to read and write; and in general are better educated than children of day-labourers, which qualifies them better for being artiſts and manufacturers. They are alſo commonly more numerous, being better nouriſhed during nonage, and better preſerved from diſeaſes. Small farms accordingly are not only favourable to population, but to the moſt uſeful population. I would not therefore indulge willingly any farm beyond a middle ſize. And to check thoſe of a larger ſize, I am clear for a tax of L. 3 or L. 4 yearly upon every farm that requires three ploughs; and ſo on according to the number. I except proprietors, who ought to be encouraged to improve their eſtates: let them employ as many ploughs as they find convenient, and not be ſubjected to any tax. If any undertaker be willing to lay out a large ſum of money upon farming, the profit of of a long leaſe will enable him to pay the tax. This tax at the ſame time may be ſo contrived, as to anſwer a valuable purpoſe, that of exciting farmers to uſe oxen inſtead of horſes; which will be done by exempting oxen-ploughs from the tax. And the undertaker mentioned will be relieved from the tax altogether, if he employ ſuch ploughs only.

THE ſize of a farm being adjuſted, next in order are its accommodations. The firſt accommodation I ſhall mention is, an acre planted with trees proper for the farm; fir in particular and aſh. The worſt ſoil will anſwer the former, and moiſt ſoil the latter. Theſe are at hand for paling and other purpoſes; and without ſuch a convenience a tenant's hands are in a meaſure bound up.

Another accommodation follows. However well adjuſted the ſize of a farm may be to the ſtrength employed upon it; yet to have a farm in perfection there ought always to be ſome by-work, about which the ſervants and horſes may be employed when the ordinary culture of the farm does not require them. There is a great difference between good and bad weather, with reſpect to expedition. The farmer muſt provide againſt the latter, by ſufficiency of men and horſes, which in the former have not full employment. Therefore, to provide againſt want of

work,

work, there ought to be a reſource, ſuch as may be taken up or laid down as occaſion offers. The carrying lime to a farm, or marl, or dung from a neighbouring town, are excellent reſources.

The chief accommodation of all is, a fruitful kitchen-garden. Formerly, oat-meal was the only food of our labouring people; and when at five or ſix ſhillings *per* boll, there could not be cheaper food. A kitchen-garden was at that time a ſort of luxury, and our ſimple peaſants had no notion of luxury. By a great advance in the price of oats and oat-meal, a kitchen-garden has become an article of œconomy; and yet ſcarce more attended to than formerly. Well dreſſed and cropped, it will afford half maintenance to a family; and yet this food would not coſt much above the third of the price of oat-meal. But the ignorant are ſlaves to cuſtom; and no value is put on a kitchen-garden at preſent, becauſe it was of no value thirty years ago.

The extent of this garden depends on the manner of cropping the farm. Where red clover is one of the crops in the rotation, a kitchen-garden need not exceed two or three acres. Where red clover enters not into the rotation, ſix acres are the ſmalleſt quantity even for a ſingle plough; becauſe it muſt yield food both for man and beaſt. Take the following rotation, which at the ſame time is capable of greater variety in a kitchen-garden of greater extent than ſix acres. Two acres are neceſſary for ſummer-food to the farm-cattle, cows, *&c.* Two acres more will be uſefully employed in cabbage, colewort, turnip, carrot, potatoes, leeks, onions turky-beans, white peaſe and other kitchen-ſtuff, for food to the family. The remaining two acres muſt be ſown with barley and red clover. To give directions for the culture of the vegetables intended for the family-uſe, would be a treatiſe on the kitchen-garden. I ſhall only ſlightly obſerve, that the early cabbage, termed *May cabbage*, may be ſet the end of February, or as ſoon after as the ſeaſon anſwers; and will be ready for eating in May and June. Scotch cabbage ought to be ſet in March; and a few ſet in June will be ready for eating next ſpring. The ſeaſon for ſowing carrots, onions, and leeks, is the ſame with that of oats. Onions require a rich ſoil; and if that be wanting, leeks may ſupply their place. Potatoes ought never to be neglected: they make a hearty food, and the cheapeſt of all: upon which account, as many ought to be raiſed as will ſerve the family the year round. Turnip is proper for cows after calving; and to turnip may be added what cabbage or colewort are ſaved from the family-uſe. This ſucculent food will produce plenty of milk for rearing calves, an article little regarded in Scotland. Careful culture may afford ſome reſidue for feeding an old cow or ox during winter; which in ſpring may be ſold to great advantage,

tage, inftead of being fold the beginning of winter for a trifle.

A kitchen-garden is always furrounded with trees, from the notion that it requires fhelter. Young plants require fhelter; but not thofe that are advanced in growth. Plants long fheltered in a kitchen-garden, are too tender for the field. And for evidence that they profper greatly by a free circulation of air, they rife in the field to a much greater fize than in a fheltered garden. When the trees that furround our kitchen-gardens are grown up, they fill the ground with their roots, and overfhadow the plants with their branches. The plants are dwarfifh even with all the dung that can be fcraped together; and yet the vegetable foil becomes deeper and deeper every year. I venture to pronounce, that it would be a confiderable improvement in hufbandry, to abandon all our kitchen-gardens, to chufe proper fpots for them well ventilated, and to make manure of the foil of the old ones. Several thoufand acres of poor land may by that means be improved. I have found great profit by that manure. I need fcarce add, that the kitchen-garden ought to be as near the farm-houfes as may be, in order to fave the carriage of dung to it, and of green food to the cattle.

There is one accommodation of a farm I relifh greatly, becaufe it has a charitable view. Why not referve work for fuperannuated labourers that requires

requires little ſtrength, cutting down weeds for example, weeding hedges, filling up ruts in a road with ſmall ſtones, tending cattle in the houſe during winter, gathering dung, and ſuch like? Half-pay would help them to live; and gentle work would keep them in ſpirits, preventing a langour that ſits down heavy on the induſtrious, when reduced to idleneſs.

It will not be an unneceſſary addition, to examine what ſort of farm is proper for a wright, a ſmith, a maſon, or a weaver. A mechanic, it is true, makes more by his art than by huſbandry-work: but if he have no farm to depend on, he muſt go far to market for proviſions, and will be ill provided after all. At any rate, he muſt have a ſmall farm for maintaining a horſe and two cows: the former for carrying timber, iron, coals, &c.; the latter for milk.

The quantity of land neceſſary for ſupplying all his wants, cannot be leſs than ſix acres where the ſoil is good, and more where it is indifferent. To prevent the expence of winter-herding, the field muſt be incloſed; but diviſions are not neceſſary. It ought to be cropped as follows: a third part in red clover, with a ſmall proportion of ryegraſs to thicken the crop: a third part in oats, turnip and potatoes; the remaining third in barley and lint; the lint to be ſown where the potatoes grew the former year.

To

To ſave the expence and trouble of carrying clover to the cattle in the houſe, and of carrying dung to the field, let a moveable trough be provided twelve feet long, the ſide ſpars eighteen inches high. Bind the horſe and two cows to ſtakes at the ſide of the trough, and let them have plenty of clover. To defend them from heat and from flies, cover them with a ſhed, the roof of which may be an old ſail, or any other coarſe cloth, ſupported upon wooden pillars. The trough and ſhed moved from place to place, will ſpread the dung equally, and ſave carriage of the clover. The wife or maid-ſervant, when ſhe milks the cows, fills the trough with clover; and the richneſs of the food will make it proper to milk them thrice a-day. The oat and barley crops need no inſtruction. The turnip is an excellent food for the cows in ſpring after calving; and there ought to be as many potatoes as to ſerve the family the year round. As for the flax, all the ground that can poſſibly be ſpared ought to be cropped with it; becauſe every perſon of the family may be profitably employed, upon dreſſing it in winter-evenings when no other work is on hand, and when otherwiſe they would be idle.

The oat and barley ſtraw, with what clover has been ſaved, will ſuffice for winter-food to the horſe and two cows. And the dung that can be thus gathered, will ſuffice for the turnip and potato

tato crops. As the lint fucceeds the potato crop, it will require no dung.

The great lofs of a fmall farm that cannot maintain a plough is, that the land is never dreffed in feafon. No neighbouring farmer will part with his plough, even for hire, till his own work be finifhed. This lofs may be prevented by placing two mechanics together. If good natured and fociable, they will go on pleafantly with their two horfes in a plough, and will have always the feafon at command for plowing. The great advantage of fuch a plan, both for frugality and profit, will fubdue even bad humour, and make the neighbours go on cordially together.

With refpect to a weaver, fome Irifh pamphlets declare againft his having any land to cultivate; for the reafon above mentioned, that there is more profit by weaving than by farming. But this reafon is not fatisfactory. Weaving, a fedentary occupation, requires at times field-labour for fake of health; and it is not confiftent with humanity, that profit only fhould be the object, without regard to health or comfort. It may at the fame time be doubted, whether upon the whole it will not be profitable to preferve a weaver a few years longer in health and vigour. Thefe confiderations muft weigh, even where there is a market at hand for provifions, which is a rare cafe in Scotland. But as a weaver has no occafion for a horfe, four acres are fufficient for his

his purpoſe, to be cropped in three diviſions as above-mentioned.

CHAP. XIV.

WHAT A CORN-FARM OUGHT TO YIELD IN RENT.

IN leaſing a farm, it appears an equitable bargain, that after a moderate allowance to the tenant for his ſkill and induſtry, and after deducting the expence of management, the remainder of the product, or its value in money, ought to go to the landlord as rent. In order then to aſcertain the rent of a farm, the particulars muſt be ſtated with which the crop ought to be burdened. This will be found not an eaſy taſk: but the following conſiderations, will pave the way to it.

The labour beſtowed on dreſſing land for a crop, is evidently a burden on the crop. But by what rule is that labour to be eſtimated? If the labour of every man and every horſe were allowed to the tenant at ſo much *per* day, it would encourage him to loiter, inſtead of being diligent: the crop would be exhauſted by the expence, and nothing be left to the landlord. The nature of the agreement between landlord and tenant, ſuggeſts a more proper method of ſtating the account. The tenant furniſhes the ſervants and cattle:

cattle; but as they labour for the landlord, the expence muſt be defrayed by him. The article for ſervants hired by the year or half-year, is ſimple: their wages muſt be ſtated as a burden on the product. The article for the labouring cattle is more intricate. Were the price of every beaſt to be ſtated, bought by the tenant during his leaſe in place of thoſe that fail, it would occaſion much trouble, and open a door to fraud. The only practicable way, is to allow the tenant a yearly ſum, for the uſe of his cattle, and for upholding them. For the former he cannot be offered leſs than five *per cent.*; becauſe that ſum he would be intitled to, upon converting his ſtocking into money. As to the latter, cattle are a periſhable commodity; and every farmer reckons upon the expence of renewing his ſtock every ten years. To anſwer this expence, the tenant ought to be allowed yearly ten *per cent.* of the value of his ſtock; which ſum in ten years amounts to no more but that value. But this is not all: a poor tenant who commonly has nothing to depend on but his ſtock, muſt be preſerved from riſk. His ſtock of labouring cattle ought to be inſured to him; and the inſurance cannot well be yearly under four or five *per cent.* Upon the whole, for furniſhing labouring cattle, and upholding them, he is well intitled to twenty *per cent.* yearly of their value; and this ſum is another deduction from the yearly product.

In

In this account muſt alſo be comprehended the farm-inſtruments, ploughs, carts, *&c.*; for theſe muſt be renewed from time to time, ſtill more frequently than the labouring cattle. But cows and calves, and other particulars that are kept for the benefit of the tenant ſolely, are not to be comprehended; becauſe they produce nothing to the landlord.

What the tenant himſelf ought to be allowed out of the product, for beſtowing his whole time upon the ſervice of his landlord, is far from being obvious. A tenant is in a higher claſs than a hired ſervant or day-labourer; and may well rank with a ſhop-keeper in a town, or a manufacturer in a village. To judge by that compariſon, his allowance ought not to be under L. 36 or L. 40 yearly; ſuppoſing the corn, one grain with another, to give ten ſhillings *per* boll, which is the price to be underſtood in every branch of this inquiry. Nor is this a high allowance, conſidering that after maintenance of the family, and education of the children, very little will be ſaved. But here nature ſtrikes out againſt this allowance. Men are ambitious of power, the lower claſſes not excepted. Every day-labourer who has ſaved a little money by penury, immediately commences farmer. He purchaſes his labouring cattle upon credit, and depends on his little ſtock for what elſe he wants. What tickles him, is not independence only, but to have the command of ſervants and horſes. Many a poor man is involved thus in difficulties, who lived more at eaſe while he was a day-labourer. One acquainted with this ſcene, would be amazed to hear, that in many places of Scotland, there are two, three, ſometimes four, tenants about a ſingle plough, in a poor farm that perhaps does not pay L. 5 yearly. Were it not the ambition of being tenants, better it would be for them to labour at 8 d. *per* day. By this ambitious propenſity in my countrymen, much leſs is made by farming, than the tenant is intitled to from the nature of his profeſſion. It is believed, that thoſe who are confined to one plough, do not at a medium clear more than L. 20 yearly; which exceeds not the wages of an ordinary mechanic, and is not far above what a day-labourer earns. The landlord indeed appears to profit by this propenſity; becauſe the leſs that is drawn of the product by the tenant, the more is left to the landlord: but the profit is only imaginary; for the landlord ſuffers in the main by the riſk he runs of weak tenants. Men of prudence, find it their intereſt to keep their farms low for the encouragement of more vigorous tenants.

However this be, the rent of ſmall farms determines the rent of every farm; for a landlord, who draws of rent L. 80 from two ſmall tenants, will not eaſily be perſuaded to ſet both farms to one tenant for L. 75. But it falls out luckily here,

here, that there is encouragement to great tenants, without encroaching on the landlord's intereſt. A leaſe of ſix ſmall farms will intitle him to the ſhare of each of the former poſſeſſors, which is L. 120, without lowering the landlord's rent one farthing.

I need ſcarce mention, that the expence of ſeed is a burden upon the annual product; as alſo the expence of reaping, threſhing, *&c.* as far as theſe articles are performed by ſtrangers hired for that purpoſe. And to ſave a vague account, a lump ſum ought to be ſtated for theſe particulars.

It falls in here to be conſidered, whether the nature of the ſoil, good or bad, make any difference upon the tenant's ſhare of the product. One thing is evident, that if he be not enabled to live by his farm, he muſt take himſelf to ſome other occupation; and to live with any degree of comfort, leſs he cannot have than L. 20 yearly, however mean the ſoil may be. On the other hand, he is intitled to no benefit from the fertility of the ſoil; becauſe it adds neither to the expence of culture nor of living. Fertility is a quality of land; and a ſubject belongs to the proprietor with all its qualities. As fertility depends not on the tenant's ſkill nor induſtry, he is intitled to no benefit from it.

For the ſame reaſon, any ſaving in the expence of culture ought to benefit the landlord only; as where,

where, by the conſtruction of a plough, two draught-horſes are ſufficient inſtead of three or four: or where oxen are uſed inſtead of horſes: the gain of the tenant is not leſſened by ſuch ſavings.

On the other hand, where a tenant, by ſuperior ſkill or extreme diligence, raiſes on an acre a buſhel more than uſual, the profit ought to be his own: it is owing to himſelf, not to the fertility of the ſoil.

Laſtly, Suppoſe a lime-quarry, or a bed of ſhell-marl, to be diſcovered within a farm, or near it, it ought to be conſidered as an article for the landlord, in giving a new leaſe. The profit ought to be his, ſtating only an allowance to the tenant for the expence he lays out upon the manure. It is in effect adding to the fertility of the ſoil; which, for the reaſon above given, ought to benefit the landlord only.

Let us illuſtrate theſe general views, by ſtating an account of particulars. Take a farm of ſixty acres; which being partly in paſture, may be managed by a ſingle plough with four horſes. I begin with computing the rent of ſuch a farm, where the product in corn and graſs is at a medium equal to the value of five bolls *per* acre, or 50 s. amounting upon the whole farm to L. 150. Add the profit of ten winterers fed with ſtraw, which may be ſtated at L. 5. The whole ſum drawn out of the land is L. 155; from which is to be deducted

deducted the tenant's ſhare, and every other article of expence: the balance is the landlord's rent. Let us enter into the ſeveral articles of deduction.

Firſt, The ſeed, which ſhall be ſtated at L. 20 only, as a part is in paſture.

Second, The fifth part, or 20 *per cent.* of the value of the labouring ſtock, which ſtock by computation is L. 74, 10 s. *. *Inde*, L. 14, 18 s.

Third, The farmer's ſhare L. 20.

Fourth, As the farmer himſelf may ſtand for one ſervant, I ſtate only the wages and maintenance of another L. 12.

Fifth, The maintenance of the four horſes L. 24.

Laſtly, The money paid for reaping, threſhing, *&c.* lumped at L. 8.

Theſe deductions amount to L. 98, 18 s. But if the land can be managed with two horſes, the deductions will amount to L. 82, 2 s. only, beſide ſaving a driver.

* Four horſes - -	L. 48	0	0
Horſe-furniture - -	2	0	0
Two ploughs - -	2	0	0
Carts and wains - -	14	0	0
Harrows and brake - -	2	0	0
Roller - - -	1	0	0
Fanner - - -	2	0	0
Forks, ſpades, ſcythes, rakes, wheelbarrows, hooks, *&c.* - -	1	10	0
Twelve harden ſacks - -	2	0	0
	74	10	0

N. B. Reparation of houses, and other small articles, are too minute to enter into a general view. But if any article be thought too high, they may serve to balance what is substracted from that article.

The account then stands thus. On the one

hand the product -	L. 155	0	0
Subtract on the other hand -	98	18	0
	56	2	0

This balance of L. 56, 2 s. is the landlord's rent.

Supposing the product to be but four bolls *per*

acre, or 40 s. ; *inde* the product	L. 125	0	0
Subtract as before - -	98	18	0
Rent	L. 26	2	0

Supposing the product to be 3½ bolls *per* acre,

or 35 s. ; *inde* - -	L. 110	0	0
Subtract - -	98	18	0
Rent	L. 11	2	0

Here an unexpected discovery is made of great importance in farming; which is, that a farm yielding no more but an average of 3½ bolls *per* acre, had better be wholly set for pasture. For supposing it in that shape to yield no more but 5 s. *per* acre, which is L. 15 for the whole, the

the clear profit is greater than when the farm is in corn; and the landlord draws more rent: he draws the whole L. 15, as land ſet in paſture is not burdened with any expence. This diſcovery may be of uſe to many a poor tenant, who labours and toils at the plough from year to year, to his own loſs. If his farm produce not more than 3½ bolls *per* acre, better abandon the plough, turn his farm into paſture, and ſit idle.

But if a tenant can reduce his labouring cattle to two horſes or two oxen, the ſaving will be ſo conſiderable as to make it his own intereſt as well as that of the landlord to continue his corn-crops. The ſaving amounts to L. 16, 16 s. yearly, not to mention the driver; which throws the balance againſt paſture no leſs than L. 12, 18 s. ſuppoſing the farm in paſture to yield but 5 s. *per* acre. This ſaving benefits the tenant during his leaſe, and benefits the landlord in giving a new leaſe. But ſuppoſing the product to be but 3 bolls *per* acre, the balance turns clearly for paſture, after every ſaving that can be made when ſuch a farm is in corn.

At the ſame time, however fertile the ſoil may be, the man who takes in leaſe a made farm, cannot expect more but to live comfortably. If his heart be ſet on wealth, it is not to be found but in land recently broken up from the ſtate of nature, where there is ſcope for great and laſting improvements.

This

This ſpeculation ought to be a ruling motive with every gentleman of a land-eſtate, to encourage improvements with all his might. The expence of culture is not leſs in a mean ſoil than in the moſt fertile; and we now ſee, that the expence of the former cuts ſo deep into the corn-product, as to leave little or nothing to the landlord. In our corn-counties, it is not difficult, nor extremely expenſive, to make the land carry two or three buſhels *per* acre more than the ordinary product; the value of which goes entirely to the landlord as rent.

Upon a review, the only doubtful article is the 5 *per cent.* ſtated for enſuring the tenant's labouring cattle. It appears to me, that a yearly ſum preciſely equivalent to the chance of loſing cattle, is not ſufficient; for if the chance go againſt the poor tenant, he is undone. Something for inſurance he ought to have; more or leſs is arbitrary. But ſuppoſing it to be 3 *per cent.* or 2 only, it will be eaſy to frame the computation upon that ſuppoſition.

PART

PART II.

THEORY OF AGRICULTURE.

THE operations of men can be eaſily traced: they are confined to the ſurface of bodies. The operations of nature, going far out of ſight, reach even elementary particles. In explaining therefore natural effects, we ought to reſt ſatisfied with the immediate cauſes, leaving the more remote to ſuperior beings. In order to unfold the theory of agriculture, the nature of plants ought to be ſtudied, their nouriſhment, their propagation: we ought to be acquainted with all the different ſoils, and in what manner they are affected by weather and climate. And yet, after all our reſearches, how imperfect remains our knowledge of theſe particulars! Fortunately, agriculture depends not much on theory. If it did, baneful it would be to the human race: ſkilful practitioners would be rare; and agriculture, upon which we depend for food, would, by frequent diſappointments, be proſecuted with little ardour. Notwithſtanding therefore that the theory of agriculture is ſtill in its infant ſtate, the practice has made conſiderable advances, eſpecially in Britain; and there are rules founded on experience, that ſeldom miſlead when applied by a ſagacious farmer. In theory, the deepeſt penetration preſerves not writers from wide differences. In practice, the ignorant only differ : ſagacious farmers generally agree ; giving allowance only for varieties in ſoil and climate.

But admitting experience to be our only ſure guide, theory however ought not to be rejected, even by a practical farmer. Man is made for knowledge ; and he has a natural curioſity to learn the reaſon of every thing. Why not indulge an appetite, that will amuſe, and may bring forth inſtruction ? In dipping into theory, a complete ſyſtem is far from my thoughts, and far above my reach. I venture only to ſelect a few particulars, that have an immediate influence on practice. Theſe will be underſtood by every gentleman who joins reading to experience ; and in doubtful caſes may help to direct his operations. I give warning before hand, that I pretend to no demonſtration. However poſitively I may happen to expreſs myſelf in the glow of compoſition, my beſt arguments are but conjectural. Thoſe that are here diſplayed appear to me highly probable ; and if they appear ſo to the reader, I can have no farther wiſh.

The ſubjects handled in this part of my work, are divided into three chapters. In the firſt are contained

contained ſome preliminary obſervations that have an immediate influence on practice. In the ſecond are handled the food of plants, and fertility of ſoil. And the third is upon the means of fertilizing ſoil.

CHAPTER I.

PRELIMINARY OBSERVATIONS.

TO be an expert farmer, it is not neceſſary that a gentleman be a profound chemiſt. There are however certain chemical principles relative to agriculture, that no farmer of education ought to be ignorant of. Such as appear the moſt neceſſary ſhall be here ſtated, beginning with elective attraction and repulſion, which make a capital article in the ſcience of agriculture as well as of chemiſtry.

1. ELECTIVE ATTRACTION AND REPULSION.

BY an inherent quality of matter, every particle of it has a tendency to unite with every other particle; and this tendency is termed *gravity*. Beſide gravity, inherent in all matter, there is in ſome bodies a peculiar tendency to unite together; acids and alkalies for example, air

air and water, clay and water. The particular bodies thus disposed to unite, may be termed *corresponding bodies ;* and as such disposition or tendency has a resemblance to choice in voluntary agents, it is termed *elective attraction.*

The power of gravity extends as far as matter exists. Elective attraction, on the contrary, is confined within a very narrow space : it operates not but between bodies in contact, or approaching nearly to it.

Even in the largest bodies, such as the sun and planets, every particle of matter operates by its power of gravity. But elective attraction has no sensible effect between large bodies, no particle operating but those in contact, or near it. It has no sensible effect therefore but between bodies that mix together. The attraction of gravity between two bodies is in the direct proportion to the quantity of their matter : elective attraction is in the inverse proportion ; which in plain language is saying, that the less bodies are, the greater is their elective attraction. Between large bodies accordingly, elective attraction in opposition to gravity is as nothing : between very small bodies in contact, or near it, it is far superior to gravity.

The power of gravity in each particle of matter corresponds to the quantity of matter in the universe : double that quantity, and you double the power

power of gravity in every particle : annihilate the half, and the power of gravity is reduced to the half. Elective attraction, on the contrary, is invariable. It can hold but a certain quantity of correſponding matter : bring more within its ſphere of attraction, it has no effect. Thus, water will hold ſalt till it be ſaturated ; or, in other words, till every particle of the water be in contact with a particle of ſalt. Add more ſalt : it is not attracted, but falls to the bottom of the veſſel. The ſame is obſervable in clay ſaturated with water : what water is added falls to the bottom.

Elective attraction between ſome correſponding bodies, is more vigorous than between others. Acids and alkalies attract each other violently, and mix intimately. Such is the caſe alſo of ſalt and water : ſalt is ſo thoroughly diſſolved in water as to vaniſh out of ſight, leaving the water tranſparent as before *. There is an elective attraction between air and water : neither of them in its

* We have Sir Iſaac Newton's authority, that the opacity of a body is owing to the reflection or refraction of the rays of light at its ſurface ; and that the particles of a body muſt be of a certain ſize to reflect or refract the rays of light. A body compoſed of ſmaller particles, is tranſparent. Water, compoſed of very ſmall particles, is tranſparent : throw ſalt into it, it maintains its tranſparency, having the power to diſſolve ſalt into very ſmall parts ; and the ſame happens with reſpect to any other ſubſtance that is diſſolvable by water into very ſmall parts.

its natural ſtate is ever found pure without the other; and yet their mixture ſeldom diſturbs their tranſparency. In the inſtances above given, elective attraction prevails over gravity, which has not power to ſeparate the heavier body from the lighter. Water and clay attract each other, but with leſs vigour: powdered clay is ſuſpended in water; but the elective attraction is not ſo ſtrong as to diſſolve the clay into its ſmalleſt parts: it continues viſible in the mixture, and makes the water turbid. Their mutual attraction yields by degrees to the repeated impulſes of gravity: the clay ſubſides, leaving the water tranſparent as originally. But each particle of clay draws along with it the particles of water with which it is in contact. And accordingly when the water is poured off, the clay remains moiſt and ſoft.

Both air and clay attract water; and when they act in oppoſition, it is of importance to know which of them prevails. Where clay is ſo wet as that many particles of the water are not reached by the attraction of the clay, ſuch looſe particles are attracted by the air without oppoſition. Even particles of water, barely within the ſphere of attraction of the clay, are drawn up by the ſuperior attraction of dry air. But the air muſt be both hot and dry, to carry off water that is in actual contact with the clay. This I conjecture never happens in pure clay, unleſs the heat be intenſe.

All

All ſorts of earth do not attract water equally: the attraction of clay is the ſtrongeſt, of ſand the weakeſt. Between theſe extremes, ſoils vary in every degree with reſpect to their power of attracting and holding water. Even clays differ. Some clays attract water vigorouſly, others leſs. This is manifeſted from the time that is taken in drying; as the clay that attracts the moſt vigorouſly will be the lateſt in parting with its water. I witneſſed an experiment of different clays put in ſhallow veſſels, and ſoaked with water. They differed in the times of drying at leaſt a fortnight. I take it for granted, that the clay which retains its water the longeſt, will turn the hardeſt after it is dry; for both effects depend on the force of the elective attraction.

It is laid down above, that elective attraction is the greateſt between the ſmalleſt bodies. Hence an important leſſon in agriculture, which may be juſtly eſteemed the corner-ſtone of the fabric, that the more pulverized earth is, the more water it will hold. A lump of dry clay immerſed in water, carries none away but what is attracted by the ſurface-particles. Pulverize this lump, and give free admiſſion to the water; let it be divided into a million of parts or into ten million: each particle however minute will hold a certain proportion of water; and this proceſs may be carried on as far as clay and water can be divided

by

by the hand of man *. It is amazing what a quantity of water may be contained in clay well pulverized, without even appearing moiſt. Shell-lime requires at leaſt its own weight of water to ſlake it perfectly; in which ſtate it is a dry powder without any appearance of moiſture. Clay well pulverized is ſimilar: every particle of it attracts a particle of water; and after perfect ſaturation, it ſtill appears dry. This I conjecture to be the propereſt condition of ground for throwing ſeed into it. Now as earth ſerves to retain moiſture, and to furniſh it gradually to its plants, the chief object of huſbandry is, by plowing and harrowing, to pulverize clay, and every other ſoil that requires it.

There is alſo obſervable an elective attraction between earth and air. Much air is found in earth, becauſe gravity concurs with the elective attraction to bring down air. But very little earth is found in air; becauſe in that caſe gravity counteracts the elective attraction.

A plant attracts air and water, and is attracted by them. The latter attraction is without effect, becauſe plants are fixed to a place. The former is clearly diſplayed by Dr Hales in his ſtatical eſſays, containing the beſt conducted experiments that

* Mr Evelyn dug a deep hole in the ground, reduced the earth to powder, and put it back into the hole. After a time, the powdered earth was found moiſt to the bottom, the ground round it remaining hard and dry.

that are known, next to thoſe of Sir Iſaac Newton upon light and colours. There is not the ſlighteſt evidence, that plants attract any dry matter, however pulverized. Set the moſt healthy vegetable in dried earth, or in duſt gathered from the highway: it dies, and the earth remains as weighty as before. It is clear then, that a plant can receive no nouriſhment but what is conveyed to it by air or water; and conſequently that nothing can ſerve as its nouriſhment but what is ſoluble in theſe elements. Earth is not ſoluble even in water: it is eaſily ſeparated from water by the force of gravity; and its particles are probably too groſs to enter with water into the mouths of a plant.

Black bodies attract and abſorb rays of the ſun. A black wall facing the ſun, is hotter even to the touch than a wall of any other colour; and hence the practice of blacking fruit-walls. Soil made black by high culture, attracts and abſorbs rays of the ſun in plenty; and turns remarkably hotter than ſoil of any other colour.

That there is a mutual attraction between particles of water, appears from the globules it forms itſelf into in falling; from the globules it forms itſelf into when dropt gently upon a dry board; and from its riſing above the brim when gently poured into a glaſs. But I am uncertain whether there be any elective attraction between particles of clay: put dry powdered clay into a veſſel, and preſs it together at pleaſure; it comes out with little or no coheſion. Water is the cement that hardens particles of clay into a ſolid lump; and it is in the propereſt ſtate for hardening, where every particle of the one is in contact with a particle of the other. Where there is more water than to admit of ſuch mutual contact, what ſuperabounds is carried off by the air in its ordinary ſtate of dryneſs. But to carry off any particle of water in contact with a particle of clay, ſtrong clay eſpecially, a very dry air is requiſite, and perhaps alſo a very hot air. The reaſon is, that water is attracted more ſtrongly by clay, than by air in its ordinary ſtate. A green turf from a moiſt ſoil, falls to pieces in handling. Let it lie a few days to dry, it becomes tough and firm. The like happens in a mixture of quick lime, ſand, and water. The water continues for a time fluid; and the maſs is ſoft and ductile. Upon evaporation of the ſuperfluous moiſture, the elective attraction operates; and the maſs turns hard like a ſtone. A mixture of clay and ſand moiſtened with water, continues long ſoft; but in time turns exceedingly hard: the pier of Eyemouth in Berwickſhire is built of a plumcake ſtone, compoſed of pebbles, clay, and ſand, cemented with water; yet no ſtone is harder, nor leſs affected with the ſea-air. Plaſter of Paris is compoſed of gypſum and water; which are mixed together to a certain conſiſtence; and the

the mafs, ftill fluid, is poured into a mould: in a few minutes it aquires a ftony hardnefs. A compofition of that kind is ufed for bridges in the ifland Minorca: no fooner is one ftone of the arch joined to another, than it bears a man to add a third. The cement here operates almoft inftantaneoufly: water operates flower in hardening clay; and ftill flower in hardening a mixture of lime and fand.

Thefe differences depend probably on the more or lefs vigour of the elective attraction. As the fuperfluous moifture evaporates, the correfponding bodies approach nearer and nearer to each other, and at laft unite in one mafs. The evaporation is flow in proportion to the vigour of the elective attraction; and the flower the evaporation is, the mafs becomes the harder. Some forts of clay attract water more vigoroufly than others; and when the fuperfluous moifture is exhaled, which is done very flowly, the mafs turns hard in proportion. Carfe-clay affords a good inftance. It is compofed of the fineft parts of natural clay, wafhed off by running water: it is depofited in flat ground where the water ftagnates; and by gradual accumulation, the ground rifes above the ftream. Carfe-lands are generally near the fea, and the reflux of the tide contributes to the effect. By the minutenefs of its parts, carfe-clay mixes fo intimately with water as to give the elective attraction its utmoft efficacy.

cacy. Thus carſe-clay, which is extremely wet in winter, becomes ſo hard in a droughty ſpring as to yield but a very ſcanty crop, unleſs the ſummer be moiſt. In a dry ſummer, fiſſures are every where ſeen in it, ſome of them ſo wide as to admit a man's hand. Common clay, compoſed of groſſer parts, never hardens ſo much. Hence it is, that of all crops beans thrive the beſt in carſe-clay: the tap root puſhes vigorouſly into the hard ſoil; and finds more water locked up there than in common clay *.

This tendency to hardneſs in clay-ſoil, is a great obſtruction to fertility; and to counteract that bad quality, I know no means more effectual than frequent plowing and harrowing. In that view partly, the harrows above deſcribed were invented: they divide the ſoil into minute parts: every part holds a particle of water ready for the nouriſhment of plants; and the ſoil at the ſame time is

* The clay in the Carſe of Gowry, of Falkirk, and of Stirling, is much of the ſame nature. When dry, it is white, leſs weighty than common clay, pure without ſand, and divides into very minute parts. The laſt-mentioned quality, which gives elective attraction its greateſt efficacy, makes it cake at the ſurface when ſtirred before winter. It makes it alſo in drought unite into very hard clods, harder than thoſe of common clay, compoſed of groſſer parts. As all clays hold water in proportion to the minuteneſs of their parts, froſt acts more vigorouſly upon carſe-clay than upon the ordinary ſort; becauſe froſt acts upon bodies in proportion to the quantity of water in them.

is kept open, inviting the roots to extend themſelves in all directions.

The time that water takes to harden clay, will explain ſeveral articles. I mention firſt an article of importance, which is the different effects of plowing clay wet or dry. The running a plough through clay ſoaked in water, produces no change. Upon evaporation of the looſer parts of the water, the vacuities left render the ſoft maſs compreſſible. To compreſs it in that ſtate would have the effect to keep in the remainder of the looſe water from evaporating, at the ſame time give the elective attraction its ſtrongeſt effect, and accelerate the hardening contrary to the very intention of plowing. Let not the plough be applied till the air has performed its part by drawing off every particle of water that is not in contact with the clay. Nick that minute for applying the plough: the clay ſtill ſoft is eaſily divided: a new ſurface is laid open to the air; and at the ſame time is preſerved free and open.

The next article I ſhall mention, is the making brick. Where air and water are brought into contact by elective attraction, the proceſs is completed: for they never harden into a ſolid body. But after clay and water are brought into contact, the proceſs goes on till they be firmly united. In the commencement of that proceſs, the air may be rendered ſo hot and dry, as to overcome the elective attraction, and ſuck up the whole moiſture,

moiſture, leaving the clay dry with little or no coheſion. Wet clay put into a hot fire, does not harden, but falls into a burnt powder. The ſuperfluous moiſture muſt be evaporated, and the coheſion be conſiderably advanced, before it can be hardened into a brick by fire. In that condition, the elective attraction prevails over the hotteſt air. Form dough into a thin cake, and lay it upon a plate of iron over a fire: the moiſture ſuddenly evaporates; and no more is left but what is barely ſufficient to keep the parts ſlightly together. Give the elective attraction time to operate: the cake turns hard like a brick *.

Nature operates by elective repulſion as well as by elective attraction; but as agriculture ſeems to depend little upon the former, I ſhall ſay but a word upon it. There is an elective repulſion in the particles of air, which gives them a tendency to recede from each other; and this operation is greatly

* With reſpect to brickmaking, where the mould is nine inches long, three broad and three thick, a brick new moulded weighs commonly eight pounds: ready for the oven, four pounds: half burnt, three pounds 12 ounces: when thoroughly burnt to be fit for uſe, three pounds eight ounces. This leſſon is of importance. If proper clay be choſen, a man cannot be deceived about the quality of his bricks. A bargain to pay for no bricks that weigh above three pounds eight ounces will enſure him. A ſmaller ſum may be agreed on for what weight a little more; which may be ſufficient for any building that is not expoſed to the external air.

greatly invigorated by heat. Heat therefore pulverizes the ſoil by rarifying the air contained in it, which moves the particles of earth out of their place. Froſt has a ſimilar effect, by rarifying the water contained in a ſoil. Black, as obſerved above, attracts and abſorbs rays of the ſun; and therefore black is the beſt colour of ſoil. The more parts clay or loam is divided into, the blacker it is. Whatever be the colour of the ſoil where potatoes are ſet, it is rendered black by that crop. A potato-crop is a powerful pulverizer: the bulbous roots ſwelling without intermiſſion, keep the ſurrounding earth in conſtant motion, and divide more effectually than a plough or a harrow. White repels the rays of the ſun; and upon that account is a bad colour for ſoil. Pulverizing by dunging, plowing and harrowing, is a ſure means to convert white ſoil into black; which is an additional motive for being diligent in theſe operations.

2. PLANTS HAVE A FACULTY TO ACCOMMODATE THEMSELVES TO THEIR SITUATION.

ALL trees are provided by nature with a tap-root, fit for piercing the hardeſt ſoil; and a tree growing in clay exerts great energy on that root. It leſſens in vigour and ſize where a tree grows in loam; and lateral roots prevail more, which are ſpread

ſpread all around for procuring food. In very light ſoil, the tap-root is very ſmall; and a tree growing in water has many roots, but not the leaſt appearance of a tap-root. Nature is wonderful in all her works. A plant here acts as if endued with the ſagacity of a thinking being: in this inſtance, and in many that will be unfolded afterward, vegetable life ſeems to be not far remote from animal life.

The conſtitution of a plant depends greatly on the ſoil it is bred in. Cuſtom becomes a ſecond nature; and it appears no leſs difficult, to tranſplant a tree from the ſoil where it was reared to an oppoſite ſoil, than to tranſplant a tree from a hot to a cold climate. However fitted by nature a tree may be for growing in a looſe ſoil; yet if planted young in a ſtiff ſoil, it acquires a conſtitution accommodated to that ſoil; and its nature is ſo far altered, as in a meaſure to diſqualify it for being tranſplanted into a looſe ſoil. Take a vegetable that has been reared in water, and plant it in a ſoil even the moſt proper for it by nature: it will infallibly die. In general, plants reared in water will not grow in earth; and plants reared in earth will not grow in water. Hence it is, that where water ſtagnates ten or twelve inches under the ſurface, the plants reared in that ground turn ſickly when their roots reach the water. Yet theſe plants would have flouriſhed in pure

pure water, had they been accuſtomed to it early.

But may not a plant acquire a conſtitution, fitting it for growing partly in earth, partly in water? Trees grow vigorouſly on the brink of a river, where ſome of the roots muſt be in water. At the ſeat of Mr Burnet of Kemnay, ten miles from Aberdeen, a kitchen-garden, a flower-garden, a wilderneſs of trees indigenous and exotic, are all in a peat-moſs, where water ſtagnates from one foot to two under the ſurface.

The ſame faculty is exerted to remedy an inconvenient ſituation. A tree that grows without ſhelter, reſiſts wind by the length of its roots: the roots of the ſame tree, are commonly much ſhorter in a ſheltered place. In the Leeward iſlands, the eaſt wind is almoſt conſtant; and the trees there, extend their roots much farther to that quarter, than to any other. A tree overtopped by neighbouring trees, directs its courſe to a ſpace that is free; and then mounts up perpendicularly according to its nature. Set a plant in a room that has no light but from a ſingle hole in the wall: inſtead of riſing perpendicularly, it directs its courſe toward the light, paſſes through the hole into open air, and then mounts upward. The power of remedying a bad ſituation, is remarkable in ſeed of every kind. A ſeed contains the plant in miniature, with a *plumula* that tends upward, and a *radicle* that tends downward. Put a ſeed into the ground with its plumula above and its radicle below as the plant grows, the former aſcends and the latter deſcends, both perpendicularly. Invert the poſition of the ſeed, the plumula ſhoots not downward, nor the radicle upward: they twiſt round the ſeed, till the former gain the open air, and the latter pierce into the ground. Providence is wonderful in every operation: were not proviſion made for the ſpringing of ſeed in every poſition, agriculture never could have made any progreſs *.

A change of conſtitution, in plants, occaſioned by their ſituation, is commonly tranſmitted to their offſpring. Plants propagated from ſeed produced in a warm ſandy ſoil, grow faſt in whatever ſoil the ſeed is ſown, and have early flowers. Plants from ſeed produced in a cold ſtiff ſoil, are late of growing, even in a warm ſoil. Plants from ſeed produced in a very rich ſoil, grow vigorouſly in a poor ſoil. Plants from ſeed of a poor ſoil, grow weakly even in the richeſt ſoil, and produce ſmall ſeed. In the rainy harveſt 1744, oats that grew in a warm light ſoil, ſprouted in the ſhock ten days more early than oats that grew in a cold ſoil; though both were produced from the ſame ſeed, and both were cut down the ſame day. Hence the advantage of changing ſeed from a warm to a cold ſoil. It may be true, that ſeed from

* See more about the powers and faculties of plants, Appendix, No. 4.

from a warm ſoil will not grow ſo quickly in a cold ſoil as in a warm ſoil; but it will always grow more quickly than ſeed from a cold ſoil. To rear trees in a middling ſoil, it is certainly right to take the young plants from a richer ſoil. But is it right to tranſplant them from a rich ſoil to one that is poor? They have, it is true, a tendency to grow vigorouſly. But will they not be dwarfiſh in the poor ſoil, which cannot afford them ſufficiency of nouriſhment to ſupport their vigour?

That a plant may change its conſtitution by being tranſplanted into a climate a little warmer or colder, is certain; and the change of conſtitution is ſtill more eaſy when the plant is raiſed from ſeed. Thus plants of one climate may, by gradual change of place in ſucceſſive generations, proſper in a very different climate. When Galen the phyſician lived, the peach was too delicate for the air of Italy. It has been creeping northward ſlowly; and, even in Britain at preſent, it is of a good flavour, if artfully cultivated. The cherry tree was brought by Lucullus from the Leſſer Aſia to Rome, as a great rarity; and now it bears good fruit even in Scotland. The bleſſings of Providence are diſtributed with an equal hand. Induſtry will remedy the natural defects of our ſoil and ſituation: are we leſs happy than thoſe who owe all to ſoil and ſituation? If wheat, if fruits, if cabbage, if collyflower, were confined

to

to their native climates, what would Britain be? Iceland would be not much inferior *.

But though a change of conſtitution is produced as far as neceſſary for accommodating a plant to a different climate, yet it is obſervable, that in other reſpects the original conſtitution remains entire. I give for one inſtance the flowering of plants. A plant tranſlated into a different climate, preſerves its original ſeaſon of flowering, unleſs prevented by ſome powerful cauſe. The climate of the ſhores of Spain and Portugal, ſuits the flowering of the *lauruſtinus* in December and January; nor is the cold of Scotland in theſe months ſufficient to deter him from his ſeaſon. I mean the milder parts; for in thoſe that are higher and more rigorous, the cold puts him paſt his ſeaſon, and prevents his ſpreading any flower till April. Dr Walker ſays, that were he to ſee a *lauruſtinus* flowering with us in winter, and had never heard of the ſhrub, he would without ſcruple pronounce it no native of this country; and that for the ſame reaſon he would deny the

* Columella, book 1. chap. 1. quotes from Saſerna the following argument to prove an alteration of climate. "Countries where neither the vine nor the olive would "grow from the ſeverity of the winter, abound now "both with wine and with oil." It is natural that this ſhould have appeared to Columella a concluſive argument; for in his days there was little experience of plants changing their nature in their gradual progreſs from hot to cold climates.

the *arbutus* to be a native of Ireland, or the whin, of Scotland. He adds pleaſantly, that the flowering of theſe ſhrubs with us, is an outlandiſh faſhion; and that no ſenſible Scotch plant will ever think of ſuch a thing.

3. CHANGE OF SEED, AND OF SPECIES.

THE reaſon for changing ſeed from a warm to a cold ſoil, is explained in the foregoing ſection. But ſkilful farmers are not ſatisfied with that ſingle change: they frequently change ſeed from a cold to a warm ſoil; and they ſeldom venture to ſow twice ſucceſſively the ſame grain in the ſame field. Such changes of ſeed, as well as of ſpecies, are common; yet I know not that the reaſon has been rightly explained by any writer. I wiſh that what follows may give ſatisfaction.

Every ſpecies of animals has a climate adapted to it, where it flouriſhes, where it grows to perfection, and where it never degenerates. Propagation will go on in a leſs proper climate; but the ſpecies degenerates, if not kept up by frequent recruits from the original climate. In that view, Arabian and Barbary horſes are from time to time imported into England. Nor is this alone ſufficient; animals procreated of the ſame breed quickly degenerate; for which reaſon, great attention is given to mix different breeds. In theſe particulars, plants reſemble animals. Britain

tain is not the native climate of melons: they degenerate quickly, if feed be not procured from the native climate. Where wheat grows naturally, feed dropping from the mother plant arrives at perfection, though neither feed nor foil be changed. But as wheat is not a native of Britain, it has a tendency to degenerate here, especially in the northern parts; and it degenerates rapidly, if the feed be fown year after year where it was produced. It is not fufficient, that the feed be taken from a different field: it ought alfo to be taken from a different foil. Nor is this all: the greateft care in changing feed will not prevent degeneracy, where the fame fpecies is fucceffively propagated in the fame field. It is accordingly a rule univerfally practifed in cropping a field, not only to bring feed from a different foil, but alfo to change the fpecies; or, in other words, to make a rotation of crops. This rule holds in barley as well as in wheat; and ftill more in red clover, which degenerates quickly when fown without intermiffion in the fame field. It is more common to fow oats after oats; and if that plant be a native of Britain, the practice may efcape cenfure, efpecially if care be taken to change the feed. White clover is a native of Britain, and requires little precaution in cropping. By Tull's mode of hufbandry, tolerable crops of wheat have been raifed in the fame field, fifteen or fixteen years fucceffively; but toward the end, the degeneracy became vifible. Artful culture will

will do much; but it is not alone ſufficient to prevail over the laws of nature. This is an objection to Tull's huſbandry, which that ingenious author did not foreſee. His mode however ought not to be totally rejected: to raiſe by artful culture, without manure, ten or twelve crops of wheat ſucceſſively in the ſame field, is a capital improvement in farms where manure is ſcarce.

The degeneracy of plants and animals in climates where they are not natives, depends on cauſes beyond the reach of human inveſtigation. But to a perſon whoſe curioſity is not boundleſs, it may be ſufficient to obſerve, that if every ſpecies of animals and plants have a climate fitted for them, there is no reaſon to expect perfection in an improper climate *.

If what is ſaid hold true, an extenſive rotation of crops in the ſame field muſt be good huſbandry. In a clay ſoil, conſtant crops of wheat after fallow without change of ſpecies, is in England not uncommon. I ſhould imagine, that in Scotland

* In the ſame field, all equally dreſſed, a firlot of Blainſly oats was ſown; and cloſs to it the like quantity of good oats produced in the farm that had not been changed for ſome years. Four bolls, two pecks, two lippies, were the product of the former. The corn weighed at the rate of 14 ſtone ten pounds *per* boll, and the ſtraw 96 ſtone. Three bolls, two firlots, one peck, were the product of the latter. The corn weighed at the rate of 13 ſtone two pounds *per* boll. And the ſtraw weighed 80 ſtone.

land wheat every other year in the ſame field, would degenerate. There is an additional reaſon againſt that practice, that it requires a larger ſtock of working cattle than a more varied rotation. A ſkilful farmer cultivates his wheat-land in October, his beans in January, his oats in March, his barley in April or May, and his turnip in June or July, all with the ſame cattle.

The particulars above ſet forth, are what I judge the moſt eſſential in the theory of agriculture, and what will be found neceſſary for underſtanding the ſubjects handled in the following chapters. Many other particulars, leſs extenſive though perhaps no leſs eſſential, are introduced where there is occaſion for them. Upon the whole, in order to eaſe the reader, I have avoided every article of theory that is not cloſely connected with practice, ſuch as a gentleman may be ignorant of, without ſuffering the imputation of being an unſkilful farmer.

CHAP. II.

Food of Plants, and fertility of Soil.

IN no branch of philoſopy are imagination and conjecture more freely indulged, than in what concerns the food of plants. Every writer

ter erects a ſyſtem : if he can give it a plauſible appearance, he inquires no farther. It never enters into his thoughts, that his ſyſtem ought to be ſubjected to the rigid touchſtone of facts and experiments : ſo grievous a torture he cannot ſubmit to. This reflection will be juſtified by what follows. And to pave the way, the method choſen by nature for feeding plants ſhall be premiſed.

Juices imbibed through the roots and leaves of a plant, are by an internal proceſs converted into ſap, which, not improperly, may be called the *chyle* of vegetables. Sap is in a continual oſcillatory motion, aſcending during the heat of day, and deſcending during the cold of night.

The ſap of a plant is nearly the ſame, in whatever ſoil the plant grows. Homberg filled a pot with earth mixed with a portion of ſaltpetre : he filled another pot with pure earth well waſhed. The creſſes that grew in theſe pots were entirely of the ſame nature, equally alkaleſcent ; what grew in the firſt pot as little acid as what grew in the ſecond. In other two pots, prepared in the ſame manner, he planted fennel, an acid plant. The difference of earth made no difference in the two plants. I advance a ſtep farther. If we can judge of ſap from what of it perſpires from the plant, it is nearly the ſame, even in different ſpecies. Dr Hales collected the liquor perſpired from trees of different kinds. It was very clear:

its ſpecific gravity was nearly the ſame with that of common water; and no difference of taſte could be perceived in the different liquors. Theſe facts are confirmed by many other experiments; all of them evincing, that however different the juices may be that are imbibed by a plant, yet that the ſap into which theſe juices are converted is the ſame, or nearly the ſame, even in plants of different ſpecies. If ſo, every plant muſt be endowed with proper powers; firſt, to imbibe juices; next, to convert into ſap the juices imbibed; and laſt, to convert that ſap into its own ſubſtance. With reſpect to the two firſt powers, all plants appear to be ſimilar. The difference of ſpecies is carried on by the laſt power only, that of converting ſap into the ſubſtance of a plant. Hence a peculiar texture, colour, ſmell, taſte, in each ſpecies. "Thus," ſays Dr Hunter of York, "a maſs of innocent earth can give life and vi- "gour to the bitter aloe and to the ſweet cane, "to the cool houſe-leek and to the fiery mu- "ſtard, to the nouriſhing wheat and to the dead- "ly night-ſhade *." In what manner or by what

* Plants are diſtributed by nature into claſſes, diſtinguiſhable by a ſimple act of viſion. Each claſs has its peculiar properties, which makes it eaſy to apply them to the purpoſes for which they are the fitteſt. Otherwiſe to attain any perfect knowledge of plants, would be an endleſs labour, and at any rate far above an ordinary mechanic. But the claſſes could never be preſerved diſtinct, if the juices imbibed by plants had any influence to vary their nature.

what means the changes mentioned are produced, will for ever remain a ſecret: they depend on energies impenetrable by the eye, and beyond the reach of experiment. Nor ought the farmer to repine at his ignorance of ſuch matters. The province of agriculture is, to cultivate ſoils in ſuch a manner as to furniſh juices in plenty: the reſt muſt be left to nature; and may ſafely be left, for ſhe never errs in her operations.

Thus prepared, we proceed to examine the moſt noted opinions concerning the food of plants. A number of writers hold, that oil and ſalt are capital ingredients in vegetable food; and that the richeſt ſoils are what contain the greateſt quantity of theſe ſubſtances. Oil and ſalt are found in vegetables. "*Ergo*," ſay theſe writers, "oil "and ſalt in the ſoil make the nouriſhment of "plants." It may as well be reaſoned, that as all animals have blood, *ergo*, blood is the nouriſhment of animals. The ſame doctrine applied to manures has led Dr Hunter, mentioned above, to propoſe a manure conſiſting chiefly of oil, termed by him the *oil compoſt*; upon which he lays great weight. Every attempt to enlighten is praiſe-worthy; but ſuch attempts ſeldom are ſucceſsful, unleſs to miſlead the credulous huſbandman. From what is ſaid above, it may be pronounced with certainty, that the oil and ſalt which enter into the compoſition of vegetables, are not imbibed from the earth or air; but are formed

formed from more ſimple materials, by the internal proceſs above mentioned. A number of plants of different kinds, may find room for growing in thirty or forty pounds of earth. Each of them has an oil and a ſalt peculiar to itſelf; though there may not be in the earth the ſmalleſt particle of either. Nay, by repeated experiments it has been found, that plants raiſed in water are compoſed of the ſame parts with thoſe raiſed in earth. Led by the above-mentioned opinion, ſeveral writers have conjectured, that clay-marl, a potent manure, muſt contain a large proportion of ſalt and oil. But it being found on trial that it contains neither, they were reduced to another conjecture, that when mixed with the ſoil it attracts ſalt and oil from the air. And now from Dr Ainſlie's accurate and elegant experiments*, that ſuppoſition appears to have as little foundation as the others mentioned.

Had due attention been given to the mouths of roots, ſo ſmall as not to be diſcernible by the naked eye, it muſt have been obvious, that oil is a ſubſtance too groſs for entering theſe orifices. Chemiſts hold, that all oils are compoſed of inflamable matter, mixed, by means of an acid, with earthy and watery particles. Nay, it is held in general, that every ſubſtance ſuſceptible of a chemical operation, is a compound; and it is natural

* Edinburgh Phyſical Eſſays, vol. 3. article 1.

natural to think ſo, for elementary particles are ſurely too minute for our handling. Now, as oil, far from being an element, is compoſed of various parts, it is certainly too groſs for the mouths of plants. Its component parts may be ſufficiently minute for admiſſion: but theſe parts are not oil; though by an internal proceſs they may be converted into oil, or converted into any other ſubſtance. Salt indeed is ſoluble in water, ſo as to become inviſible; and with water conſequently it may be imbibed by plants. But ſalt is too acrid to be a nouriſhment for plants; and if imbibed in any quantity, will be deſtructive.

I have given the ſtricteſt attention to this doctrine, in order to put the ſpeculative farmer on his guard. If he aſſent to what is here delivered, it may ſave him much time, that would be loſt in peruſing certain huſbandry-books, and much labour in proſecuting idle experiments. And here I muſt ſay again, for it cannot be too often ſaid, that the province of agriculture, is to prepare the ſoil for yielding plenty of juices, leaving the reſt to nature.

Tull is one of the boldeſt theoriſts that have come under my inſpection. He pronounces without heſitation, that all plants live on the ſame food or *pabulum*, which he ſays is pulverized earth; and upon that foundation, he pretends to raiſe perpetual crops of wheat in the ſame field, by by the plough alone, without manure. This indeed would reſtore the golden age of eaſe and indolence; as there is no ſoil ſo barren, but what may readily be pulverized. Tull was a man of genius, but miſerably defective in principles. Plants imbibe water plentifully at the leaves, bark, and roots; and with water they imbibe whatever is diſſolved in it. But earth is not ſoluble in water; and accordingly, by an experiment of Van Helmont it appears, that earth makes no part of the food of plants. He put into a veſſel two hundred pounds of dry earth; which he moiſtened with rain-water, and planted in it a cutting of willow, weighing five pounds. The mouth of the veſſel, to keep out duſt, was covered with a tin-plate, having many ſmall holes, through which rain or diſtilled water was poured from time to time, for keeping the earth moiſt. The willow weighed at the end of five years a hundred and ſixty-nine pounds and about three ounces; and the weight would have been much greater, had the leaves that fell the firſt four years been computed. At the end of the fifth year, the earth was taken out of the veſſel; and, when dried, was found to have loſt none of the original weight excepting two ounces. Mr Boyle made a ſimilar experiment with gourds, the reſult of which was the ſame *. Tull's ſyſtem then is ſingularly unlucky:

* From experiments made by Dr Woodward, he affirms, that earth is imbibed with water in a conſiderable quantity

lucky: of all fubftances, earth appears to be the leaft fitted for nourifhing plants; and from Van Helmont's experiment it is clear, that if at all it enter the mouths of plants, the quantity is inconfiderable: of the great quantity of earth, two ounces only were loft; and fuppofing thefe two ounces to have been diffolved in the water and fucked in by the plant, it was next to nothing, confidering the weight of the whole plant, which was a hundred and fixty-nine pounds.

Other writers, more cautious, avoid fpecifying any particular fubftance as the food of plants: but hold, that every fpecies requires a peculiar nourifhment; and that roots imbibe thofe jucies only which are fitted for nourifhing the plant. The refutation of this hypothefis will not require many words. It is fufficient to obferve, that it is refuted by unconteftable experiments. Plants take

quantity (Tranfactions of the Royal Society, ann. 1699). In two glafs phials full of water, he put an equal quantity of garden-mould. In one of them a plant of mint was fet, and the earth after a time was fenfibly diminifhed; but not in the other where no plant was fet. The doctor did not advert, that garden mould, which perhaps for a century had been regularly dunged, muft be replete with animal and vegetable particles that are foluble in water, and with it are imbibed by plants. By the extraction of fuch particles it is no wonder that the earth put into the phial was diminifhed. But fuch particles are not earth; and therefore the Doctor's experiments contradict not thofe of Van Helmont and Mr Boyle.

take in with air and water whatever is diſſolved in them, without diſtinguiſhing the ſalutary from the noxious : there is not the leaſt appearance in any plant of a choice. Barley has been poiſoned with brimſtone, and mint with ſalt water. Mr George Bell ſtudent of phyſic in the college of Edinburgh, an ingenious young gentleman, made the following experiments, which he obligingly imparted to me. A number of Jeruſalem artichokes were ſet in pots filled with pure ſand. One plant was kept as a ſtandard, being nouriſhed with common water only. Other plants of the ſame kind, were nouriſhed with water in which ſalt of tartar, a fixed alkali, was diſſolved. Theſe grew more vigorouſly than the ſtandard plant. But by reiterated waterings, there came to be ſuch an accumulation of the fixed alkali among the ſand, as to make the plants decay, and at laſt to die. Some plants were nouriſhed with water, in which ſal ammoniac, a volatile alkali, was diſſolved. Theſe grew alſo well for ſome time ; but, like the former, were deſtroyed by the frequent reiteration of it. Weak lime-water promoted the growth of its plants more than common water. But water completely ſaturated with quick-lime, proved more noxious than that which contained a ſolution of fixed alkali ; though leſs than that which contained a ſolution of volatile alkali. Hence appears the hurt of overdoſing a field with quick-lime. Urine promoted long the

the growth of plants; and the moſt putrid appeared to have the ſtrongeſt effect: but at laſt, it totally deſtroyed them. Water impregnated with putrid animal and vegetable ſubſtances, did more effectually promote the growth of plants than any other ſolution; and in every ſtage of the progreſs appeared to be ſalutary.

Rotation of crops in the ſame field, univerſally practiſed, is what probably has promoted the opinion of a ſpecific nouriſhment. It has been urged, that if all plants live on the ſame food, the ſoil muſt be exhauſted by a ſucceſſion of different plants, as much as by a ſucceſſion of the ſame plant. This argument for a ſpecific nouriſhment has a formidable appearance; and in order to obviate it, I found it neceſſary to give peculiar attention to that branch of huſbandry, which is done above *. It is there made evident, that change of ſpecies is neceſſary, not for food, but for preventing degeneracy. It is not want of food that makes a horſe degenerate in Britain; and as little want of food that makes wheat degenerate, where ſown without intermiſſion in the ſame field. Plants native to Britain never degenerate, though always growing in the ſame ſpot; white clover, for example, nettles, broom, whins, ruſhes, couch-graſs, &c. &c. A bull-dog never degenerates in Britain.

* Part 2. chap. 1. § 3.

Animals from their food are divided into two kinds, carnivorous and graminivorous. But I discover no such distinction among plants: they imbibe indifferently whatever is dissolved in water. And the plan of nature appears to be what follows. Certain substances were originally provided for their food. It is highly probable, that a quantity was lodged on the surface of this earth, for nourishing the first plants: whence the fertility of virgin soils, such as have never been cultivated. This matter dissolved in water and imbibed by plants, is communicated to animals that feed on plants; and is again set free by the death and putrefaction of these animals. The more volatile parts are attracted by the air: some are sucked in with air at the leaves of plants: some are washed down to the ground by rain; and with it are sucked in at the roots. The less volatile parts, which the air does not attract, are dissolved in water, and with it are also sucked in. And thus the process is continued without end. This doctrine of a common nourishment, is firmly supported by the following facts. First, Plants of different kinds growing in the same spot, rob and starve each other; which could not be, if each drew from the soil a separate nourishment. Second, Grafting and inoculating demonstrate a common nourishment. If the roots of the stock imbibe those jucies only that are proper for its own nourishment, the grafted plant must starve.

The

The juices imbibed by the former, nouriſh both; and theſe juices are by each plant converted firſt into ſap, and then into its own ſubſtance. Third, Dung of putrefied vegetables, of whatever kind they be, is one homogeneous ſubſtance; and yet vegetable dung prepares the ſoil equally for every ſort of plant. I add a conſideration of a kind that to me is always perſuaſive: a common nouriſhment is not only a more ſimple, but a more wiſe diſpenſation of Providence than a peculiar food for each ſpecies: every plant grows not every where; and if each ſpecies required a peculiar food, a vaſt ſtock of vegetable food would remain unuſed; which is not conformable to the frugality of nature, nor to the wiſdom of Providence, which makes nothing in vain.

But though all plants imbibe indifferently every ſubſtance that is diſſolved in water; it follows not, that every ſuch ſubſtance, even where innocent, is equally nouriſhing. Some ſubſtances may be proper nouriſhment, ſome not: and it may rationally be ſuppoſed, that the latter is thrown off as excrement. Why may there not be a reſemblance in this particular, between plants and graminivorous animals? A horſe, an ox, a ſheep, a goat, live all of them on graſs; but each of them have favourite graſſes, which they prefer before other kinds. This will ſo far juſtify the notion of a ſpecific nouriſhment. Let experiments be made, to try what is the moſt ſalutary food

food for plants. A few experiments of that kind are mentioned above; but to give ſatisfaction, they ought to be multiplied and extended to plants of different kinds.

There is not an opinion more generally adopted, than the following concerning agriculture, whoever was the author, That fertility of ſoil depends on the quantity of nutritive matter in it, whether ſpecific or common; that when the quantity contained in any field is exhauſted by cropping, it is reſtored by the plough, by dung, or by other manure; and that to reſtore an exhauſted field by ſuch means, is the ſole object of agriculture. This opinion has a fair appearance: nor did I ever entertain a doubt about it, till the following conſiderations happened to occur. I do not much reliſh the notion, that the number of plants growing at any time on this globe, muſt be limited by the quantity of matter created originally for their nouriſhment, nor that the quantity of graminivorous animals muſt be alſo ſo limited; and yet this muſt neceſſarily follow, if plants have no other food but what was thus originally provided for them. But ſuppoſing this conſideration not to weigh with others as with me, there are other conſiderations that cannot fail to make an impreſſion. Some countries produce corn and cattle, not only for the inhabitants, but for exportation. According to the eſtabliſhed opinion, theſe countries muſt long ago have

have been reduced to abſolute barrenneſs. Egypt and Sicily were of old the granaries of Italy; and vaſt quantities of vegetable food, converted into corn, were annually exported from theſe countries never to return. Yet we do not find that they are leſs prolific than formerly. Sicily at preſent does not conſume at home above the ſeventh part of its wheat; the remainder is exported; and yet not the leaſt ſymptom of approaching barrenneſs. Conſider the endleſs quantity of beef exported every year from Cork in Ireland: whatever quantity of vegetable food may originally have been ſtored up in that part of the iſland, it muſt long ago have been totally exhauſted. I urge another objection more general. Where-ever burying under ground is the practice, the vegetable food contained in the bodies of human beings is totally loſt, not to mention thoſe who periſh at ſea. At that rate, there is a gradual diminution of vegetable food, ſo as that in time the whole muſt be exhauſted. I add a fact to convince any thinking perſon, that plants muſt be provided with ſome food beſide that originally created. In Scotland, there are fields that paſt memory have carried ſucceſſive crops of wheat, peaſe, barley, oats, without a fallow, and without manure. And that there are ſuch fields in England and elſewhere, it is not to be doubted. A field of nine or ten acres on the river Carron, is ſtill more extraordinary. Upon on it I ſaw a good crop of oats almoſt ripe ; and by information it was the hundred and third crop of oats without intermiſſion, and without manure, as far as was known. Now, whatever be the nature of ſuch a ſoil, its unremitted fertility cannot be accounted for, from any ſuppoſed quantity of vegetable food originally accumulated in it. It is eaſy by manure to make a ſoil too rich for corn : it vegetates without end, and the ſeed has not, before winter, time to ripen. But ſuppoſing the richeſt ſoil to be proper for corn ; yet the vegetable food it contains, however great the quantity, muſt in time be exhauſted by cropping. Some other proviſion therefore muſt be made by nature for the nouriſhment of plants, beſide the vegetable food originally created.

Immenſe is the quantity of corn and ſtraw, that during a century is produced in a ſoil perpetually fertile. It is a puzzling queſtion, whence proceeds ſuch a quantity of matter ; for a new creation cannot be admitted. A perpetual effect muſt have a perpetual cauſe : the ſoil muſt receive additions without end, to reſtore what is taken away in corn without end. I am aware, that the ſmalleſt portion of matter may by diviſion be made to occupy ſpace without bounds. But obſerve, that the difficulty ariſes from weight, not from bulk. Corn is a weighty ſubſtance ; and the corn produced on this globe from the beginning, muſt amount to a weight above computation :

putation: the ſmall portion reſtored to the ground in manure, is a mere atom in compariſon. I have endeavoured above to evince, that earth is not converted into corn: and here is an additional proof; for ſuch converſion would exhibit a very new ſcene: inſtead of the hills ſinking down ſlowly into the vallies, the vallies would ſink rapidly down from the hills. A perpetual effect, I have obſerved, muſt have a perpetual cauſe: to preſerve a ſoil perpetually fertile, there muſt be a continual influx of vegetable food, to ſupply what is taken away by cropping. Whence comes that vegetable food? where is it ſtored up?

Air and water contribute to vegetation: let us try to build on that foundation. Suppoſing air and water to be the food of plants, not meaning to exclude what may be diſſolved in them, the difficulty vaniſhes, as air and water are inexhauſtible. And why may not that ſuppoſition hold in fact? I begin with air. Many plants grow to perfection, without having any nouriſhment that can be diſcovered but air only. The houſe-leek grows from choice on a dry mud wall, which ſurely affords no nouriſhment. A ſpecies of the ſedum, requiring a hot-houſe in winter, is never watered. The wall-flower grows luxuriantly in the ſeams between large ſquare ſtones in old buildings, from which all moiſture is excluded but what is in the air. Various kinds of moſs

grow

grow upon hard ſtones, where they can have no nouriſhment but from the air. It is an univerſal opinion, that leguminous plants, before they ſeed, draw moſt of their nouriſhment from the air. Conformable to that opinion, Dr Hales, in his curious Statical Eſſays, has made it evident, that every vegetable contains a quantity of air, which adds to the weight as well as to the bulk. The fixed air in a green pea, makes no leſs than a third part of the weight. In wood, however old or dry, air is found, very obſervable when ſet looſe by fire. Here is one inexhauſtible ſtore of matter for compoſing plants.

Water is another inexhauſtable ſtore. A plant regularly watered, will grow vigorouſly in the moſt barren ſoil, even in dead ſand. In Perſia, very little rain falls during ſummer, and the land is burnt up; not a pile of graſs to be ſeen. But plants there regularly watered, grow exceſſively. There are many experiments of plants ſet in glaſſes upon moſs or ſpunge, which grow well when watered. Some cotton was ſpread on water in a phial: a pea dropped on it ſprung, and puſhed roots through the cotton into the water. The plant grew vigorouſly, and bore large pods full of ripe ſeed. There is a noted experiment of an oak growing in pure water to the height of eighteen feet. Water is attracted by vegetables of every kind; and is ſucked in at the roots, at the leaves, and even at the bark.

The

The quantity imbibed during a ſpring and a ſummer, is amazing. The quantity exhaled every day, is accurately meaſured in the Statical Eſſays mentioned; which muſt be leſs than what is imbibed, becauſe plants do not throw off all they imbibe, part being converted into their ſubſtance, and adding to their bulk and weight*. The drieſt wood accordingly yields, by diſtillation, a large proportion of water. A ſtream occupies the loweſt ground, without regard to ſoil; and yet the graſs on its borders is always more rich and verdant than at a diſtance. A tree grows no where more vigorouſly than at the ſide of a brook. About a large ſtone fixed in the ground, the graſs is generally the beſt in the field: for what other reaſon, than that the rain which falls on the ſtone runs off to the ſides? I do not ſay, that the heat of the ſtone during ſummer may not contribute ſomewhat. The north ſide of a hill, is obſerved to be commonly better ſoil than the ſouth ſide: if there be truth in the obſervation, it muſt proceed from moiſture, leſs being evaporated from the former than from the latter. Black ſolid peat-

 moſs

* The emiſſion of water from plants, is the occaſion that a country abounding with trees, is more ſubject to damps, humid air, and frequent rain, than a bare country where no trees grow. The exceſſive moiſture of the American air was a great annoyance to our firſt ſettlers; but the air became more dry and the weather more conſtant, as the ground was cleared of trees.

mofs retains moifture like a fpunge: trees grow vigoroufly in it, provided they be fheltered from wind; for their roots cannot refift wind in a foil fo tender and loofe. On the other hand, no foil is more barren than a gravelly or fandy moor that holds no water; upon which the foot makes no impreffion, not even after a heavy fhower. I am bufy at prefent in cultivating a moor of that kind, upon which are fcattered fome dwarfifh plants of heath and bent, that leave half of the furface bare. I judged that manure would not anfwer, till the field fhould be made to retain moifture; for which reafon I incorporated with it a quantity of foft fpungy earth. I added lime and dung; and now it carries a rich crop of turnip and cabbage. The alteration of the foil is obvious to the eye; and alfo to feeling, as the foot dips in it after every fhower. Lord Bacon long ago gave his opinion, that for nourifhing vegetables, water is almoft all in all; and that the earth ferves but to keep the plant upright, and to preferve it from too much heat or too much cold.

To fupply the endlefs quantity of moifture neceffary for vegetation, nature has made ample provifion. The continual circulation of water upon this globe, from its furface to the atmofphere, and down again to the furface, is juftly admired for the fimplicity of its caufe, no lefs than for its bountiful effects. An elective attraction between air and water, is the *primum mobile* of thefe effects.

fects *. Water is eight hundered times heavier than air; and yet by that single power, an immense quantity of water is suspended in air; and falls down from time to time in rain, dew, and snow, impregnating the earth with moisture. Dr Hales, in a dry July, dug up a cubic foot of brick-earth, weighing one hundred and four pounds, which contained six pounds and eleven ounces of water. Under the former he dug up another cubic foot, weighing one hundred and six pounds and six ounces, which contained ten pounds of water. Under this he dug up a third cubic foot, weighing one hundred and eleven pounds and one third, which contained eight pounds and eight ounces of water. Here is a considerable stock of moisture, sufficient without rain to afford vegetable nourishment several weeks; not to mention what may be attracted from below, by the upper stratum when its moisture is exhausted. Evaporation goes on so rapidly between the tropics, that to preserve plants alive, moisture must be attracted from below: for, as mentioned in the first chapter of this part, there is an elective attraction between earth and water; and where a portion of earth is saturated with water, it readily yields its superfluous water to a dry body in contact with it. This ascent of moisture is promoted by the heat of the sun, which

* Edinburgh Physical Essays, vol. 3. art. 4.

which pierces deeper into the earth than two feet, according to experiments made by Dr Hales. Were not plants thus ſupplied with moiſture in the torrid zone, where no rain falls for many months, they would be deſtroyed by the ſcorching heat of the ſun. The evening dew that falls in a hot ſummer, is ſucked up the following day, without ever ſinking to the roots of plants.

But though air and water are made by nature the conſtant and inexhauſtible food of plants, there ſeems to be little doubt but that this food may be inriched by various ſubſtances diſſolved in them. We have ſeen above, that water impregnated with rotten animal and vegetable ſubſtances, makes rich nouriſhment for plants; and from experiments, other ſubſtances may probably be diſcovered, equally efficacious. Plain water may be ſufficient for the ſtem, branches, and other groſs parts; but we have reaſon to think, that richer nutritive matter is neceſſary for perfecting the ſeed. Hence the imperfection of ſeed in a rainy year, where the rich matter bears no proportion to the quantity of water that paſſes through a plant.

Moderate rain in a kindly ſeaſon, warms tilled land, and produces a ſlight fermentation. It is here as in a dunghill: a very ſmall quantity of moiſture has ſcarce any effect: a great quantity chills the ground, inſtead of warming it. Different plants require different quantities of moiſture.

Graſs

Graſs is benefited by all it receives; provided the moiſture exceed not ſo much, as to chill the ground and the roots of the plants *. So far corn reſembles graſs, as to be ſtunted by lack of moiſture, and conſequently to blanch early. The ſeed has more than time to ripen; but it is lank and ill filled. Corn differs in being hurt by much moiſture: it vegetates continually; and winter comes on before the ſeed begins to ripen. Holland is a moiſt country: there is ſcarce a foot of dry ground in it. Trees, graſs, and vegetables, grow there luxuriantly: but its fruits ſeldom ripen: and where ripe, have little taſte.

Thus from air and water, with what they contain, there is an inexhauſtible ſupply of vegetable nouriſhment, which fairly accounts for the immenſe quantity of corn that is annually produced.

If water be the chief food of plants, there never can be a large tree but adjacent to water running either above or under the ſurface. The experiments of Dr Hales make it appear, that plants perſpire greatly; and the perſpiration of a large

* A graſs plant cannot retain ſo much as to hurt it: whatever is imbibed more than ſufficient for nouriſhment, perſpires at the leaves. There is a conſiderable latitude in the quantity of healthy perſpiration; which, in the ſun-flower, by an experiment of Dr Hales, appears to be from ſixteen to twenty-eight ounces in twelve hours day. And he adds, that the more it was watered the more it perſpired.

large ſpreading oak muſt be very great. Part of this perſpiration muſt be ſupplied by a running ſtream; for all the rain that falls within the circumference of a tree, is not ſufficient.

The ſun joins with air and water in nouriſhing plants. The green colour of plants is occaſioned by an oily ſubſtance, which can be ſeparated by a chemical operation; and that oily ſubſtance is owing to the ſun, for no plant is green where the ſun is excluded. The ſun therefore contributes to advance plants to perfection. And it is one of the properties of leguminous plants, that their broad leaves abſorb more of the ſun's rays, than the narrow leaves of culmiferous plants.

According to the foregoing theory, the only uſe of a ſoil, is to fix the roots of plants, and to hold water for nouriſhment. But at that rate, where lies the difference between a rich and a poor ſoil. This globe is ſurrounded with air, and rains pays not homage to one field in preference to another? The ſolution of this queſtion will, if I be not groſsly miſtaken, confirm the foregoing theory, and evince that it is founded on truth. Soils originally may have been very different with reſpect to fertility, ſuppoſing vegetable food to have been unequally diſtributed by the hand of nature. A virgin ſoil may be extremely rich; witneſs the ſurpriſing fertility of America, when agriculture was introduced there. But cultivated grounds muſt long ago have been deprived

deprived of that original food, in the courfe of cropping; after which, it does not enter into my conception, what other circumftance can remain to make a foil fertile, but the holding water in fufficient quantity for its plants. A clay foil holds a great quantity; a fandy foil, very little. Some foils there are fingularly retentive of moifture; and that quality makes them long of drying: fuch foils are favourable to vegetation; for though they refift drought, they yield to the attraction of plants *. Other foils are very little retentive of moifture: they dry in an inftant, and the nourifhment they can afford is very fcanty. Here the myftery is unfolded. The richeft foil is what gives the greateft refiftance to a drying air, and at the fame time furnifhes to its plants their proper quantity of moifture. I have a thorough conviction, that this property belongs to a foil perpetually fertile: and it is to me a ftrong confirmation of the prefent theory, that I cannot form even the flighteft conception, how perpetual fertility can otherwife be accounted for; and as little can I form a conception, how otherwife countries, like Poland or Livonia, out of which great cargoes are annually exported of corn and flaxfeed, fhould fuffer no diminution of fertility.

To

* May it not be thought, that the quantity of moifture which gives to a foil its higheft fermentation, is at the fame time the fitteft for perfecting feed. However pleafing this conjecture may be, I dare not vouch it for truth: it muft be left to experiment.

To recruit with vegetable food a ſoil impoveriſhed by cropping, has hitherto been held the only object of agriculture. But here opens a grander object, worthy to employ our keeneſt induſtry, that of making a ſoil perpetually fertile. Such ſoils actually exiſt: and why ſhould it be thought, that imitation here is above the reach of art? Many are the inſtances of nature being imitated with ſucceſs: let us not deſpair while any hope remains; for invention never was exerted upon a ſubject of greater utility. The attempt may ſuggeſt proper experiments: it may open new views; and if we fail in equalling nature, may we not however hope to approach it? A ſoil perpetually fertile, muſt be endowed with a power to retain moiſture ſufficient for its plants; and at the ſame time muſt be of a nature that does not harden by moiſture. Calcarious earth promiſes to anſwer both ends: it prevents a ſoil from being hardened by water; and it may probably alſo envigorate its retentive faculty. A field that got a ſufficient doſe of clay-marl, carried above thirty ſucceſſive rich crops, without either dung or fallow. Doth not a ſoil ſo meliorated draw near to one perpetually fertile? Near the eaſt ſide of Fife, the coaſt for a mile inward is covered with ſea ſand, a foot deep or ſo; which is extremely fertile, by a mixture of ſea ſhells, reduced to powder by attrition. The powdered ſhells, being the ſame with ſhell-marl, make the ſand retentive of moiſture;

and

and yet no quantity of moiſture will unite the ſand into a ſolid body. A ſoil ſo mixed ſeems to be not far diſtant from one perpetually fertile. Theſe, it is true, are at beſt but faint eſſays; but what will not perſeverance accompliſh in a good cauſe?

A ſoil is denominated fertile, that affords plenty of nouriſhment to its plants; and accordingly it is ſuch a ſoil only that has been the ſubject of the foregoing inveſtigation. Plants that live moſtly on air, require not ſuch a ſoil; witneſs the houſe-leek mentioned above. Juniper thrives on the top of a mountain, in the pooreſt and drieſt ſoil; becauſe, in that ſituation, it is ſufficiently ſupplied with moiſture from the air. If planted in a dry climate, it requires a moiſt ſoil, to ſupply the want of a moiſt atmoſphere. The nature of the yew is the ſame. Planted in a valley, it requires a damp ſoil. It will grow in the drieſt ſoil, as in the cleft of a rock, but then it muſt be at ſuch a height, as to enjoy a humid air. Theſe belong to the claſs of plants that affect to grow in a ſoil commonly reckoned barren. A ſoil that a mere farmer would pronounce barren, is for many plants excellent. How otherwiſe could the earth be every where cloathed with beauty? How great muſt be the diverſity of ſoil, climate, and ſituation, that can raiſe to perfection above 20,000 kinds of plants, to each of which is requiſite ſome peculiarity of ſoil, of climate, or of ſituation?

Whether the theory here exhibited will occasion any material alteration in the practice of agriculture, must be left to the discovery of time. Of one thing I am firmly convinced, that the instructions above delivered, are in every particular conformable to that theory. Take the following slight specimen. · Plants, like animals, cannot live long in the same air: a circulation is to both equally requisite. For that reason, the growth of plants under cover, is slow; and seed springs slowly in stagnated air. Conformable to this observation, a free circulation of air for corn is warmly recommended; and there is a caveat against small inclosures surrounded with strips of planting, because they occasion a stagnation of air. Impure air imbibed, renders a plant unhealthy: the grain has a bad taste, and tends to generate diseases; which holds remarkably in a fruit-orchard, if the trees be crouded, and the walls high. Even grass, where the air stagnates, is unpalatable and unwholesome. Next, with respect to moisture. No branch of husbandry is more sedulously inculcated, than that of dividing and pulverizing earth by the plough, by the brake, by harrows, and by manure: it is the very life of agriculture. And from what is laid down above, it appears, that fertility depends greatly on that practice *: it increases the capacity of soil to contain water: it envigorates its retentive power; and it prevents the soil from hardening: to these ends

* Part 2. ch. 1. sect. 1.

ends manure greatly contributes; and in that view it makes the ſubject of the following chapter.

To aid the fertility of ſoil, the pickling ſeed has been much practiſed. We liſten readily to the marvelous, eſpecially where any great advantage is promiſed. The boaſted effects of the Abbé de Valemont's prolific liquor, found many believers, by which vaſt crops were to be reaped, without manure, and almoſt without plowing. (See Du Hamel's treatiſe on the culture of land, vol. 6.) And the Baron de Haac's powder, is at preſent no leſs ſucceſsful in England. The credulity of farmers might in ſome meaſure be excuſable, were ſuch bold pretenſions within the verge of poſſibility. In every ſeed there is an embryo plant; and the reſt of the ſeed ſerves to feed that plant, till it acquire roots for drawing its nouriſhment from the ſoil. The pulp is thus exhauſted, and there remains only the uſeleſs huſk. What advantage then can be derived from a prolific liquor or powder? It may poſſibly render the pulp fitter to feed the young plant, till it ſtrike root. That it can have no other effect is evident; firſt, becauſe it is exhauſted with the ſeed; and, next, that ſuppoſing any of it to remain, it can be of no benefit to roots that are ſpread an inch, two, or three, from the place where the ſeed was laid. Yet books of agriculture are ſtuffed with ſuch receipts.

I close this chapter with a reflection of the justly-esteemed Dr Hales. "Though I am sensible, "that from experience chiefly we are to expect "the most certain rules of practice; yet the "likeliest method for making the most judicious "observations, and for improving any art, is to "get the best insight we can into the nature and "properties of what we are desirous to cultivate "and improve."

CHAP. III.

MEANS OF FERTILIZING SOILS.

AS these means are plowing and manuring, they shall be treated in their order.

1. PLOWING.

THERE are mutual connections between man and the ground he treads on, that fit them for each other. The dry part of this globe, is every where covered with a *stratum* of earth, producing vegetables for the nourishment of man and of other animals. Some *strata* there are, so barren as not to bear vegetables; and some vegetables there are, that afford no nourishment: but both are rare, and intended probably for other purposes.

This

This *ſtratum* is commonly ſufficiently deep for a free courſe to the roots of plants : or it may be made ſo by art, it being one of the many purpoſes of agriculture to deepen a ſhallow ſoil. A deep ſoil beſide giving free courſe to roots, retains much water to nouriſh them. In Scotland, partly from ignorance, partly from the weakneſs of labouring cattle, it is the general practice to plow with a ſhallow furrow, commonly under four inches ; and hitherto the progreſs toward a better mode has been ſlow. It is never difficult to invent reaſons for juſtifying what we are accuſtomed to. " If we plow deep, we are afraid " of till." And what is till ? Where ground is ſtiff, rain ſettling at the bottom of the furrow cements the earth under it, which in time is hardened to a ſtone; and it is this hardened earth which is named till. The earth is hardened as far as the water penetrates, which may be one or two inches ; but after till is formed, every drop of rain reſts upon it without making any impreſſion. To ſubdue till is an important object ; and luckily the undertaking is ſeldom difficult : a ſtrong plough, raiſing it to the ſurface, lays it open to the ſun, air, and froſt, which reſtore it to its original ſtate. One precaution is neceſſary. Certain earths, as hinted above, are averſe to vegetation. Theſe muſt be avoided, however ſhallow the ſoil be: but as ſuch earths are rare,

they

they ought not to be an excuſe for ſhallow plowing.

The advantages of deep plowing, are manifold. In the firſt place, roots extend far where they meet no reſiſtance; and the growth above the ſurface correſponds to that below: roots cramped in a ſhallow ſoil, are dwarfiſh; and conſequently ſo is the tree. Dr Hales juſtly obſerves, that the greater proportion the ſurface of the roots bears to that of the tree, the greater is the vigour of the tree, and the better able to reſiſt the attacks of an unkindly ſeaſon.

In the next place, a deep ſoil affords not only ſpace for roots, but holds a due proportion of water for nouriſhing the plant. If more rain fall than the ſoil can retain, it deſcends to the bottom of the furrow, where it lies lower than the roots, or but barely touches their extremities: the ſeaſon muſt be very wet, that raiſes the water ſo high as to do much damage. The diſadvantage of a ſhallow ſoil is in that reſpect very great. Roots accuſtomed to earth are unqualified to grow in water: they ſuffer when ſoaked in water; which muſt often happen in a ſhallow ſoil, and is viſible above ground by a ſickly yellow hue *. Upon that account, Miller enters a caveat againſt overwatering tranſplanted trees: "it rots the young "fibres," ſays he, "as faſt as they grow." Dr Hales

* Part 2. chap. 1. ſect. 2.

Hales gives the ſame leſſion. There is another diſadvantage of a ſhallow ſoil: the water lodges ſo near the ſurface, as ſoon to be exhaled in dry weather. Thus, the exceſſes of moiſture and of drought, are both of them incident to a ſhallow ſoil. In a deep ſoil, water lodged at the bottom of the furrow, is a reſervoir, which is not exhauſted but by long drought. Laſtly, a deep ſoil affords ſpace for placing the ſeed ſo, as that the roots may ſpread every way: in a ſhallow ſoil, if the ſeed be laid ſo deep as to be ſufficiently covered, it approaches the hard bottom, unkindly to tender roots.

So much for a deep ſoil. I proceed to other advantages of plowing. Stiff ſoil is not penetrable by water: looſe ſoil does not retain it. Plowing diminiſhes the tenacity of the former, and opens it to receive water: it makes the latter more compact, and increaſes its power of retaining water. Some earths fill not the hole out of which they were dug: ſome do more than fill it. Poroſity occaſions the former: the pores are diminiſhed by handling, which makes it more compact. Solidity occaſions the latter: clay ſwells by ſtirring; and continues ſo, till its former ſolidity be reſtored by the power of gravity *.

Another

* This experiment ought to be tried upon ground that has reſted many years. Among the cauſes of poroſity, one is, the great number of inſects that have their dwelling

Another advantage of plowing regards clay chiefly, which by moisture turns hard if not duly stirred. This is an important article. Sand has no cohesion; and dry clay very little, if any *. It is water that cements clay; and in plowing makes it rise in lumps or clods, great or small in proportion to the degree of cohesion. Plowing prevents water from binding a clay-soil: the superfluous moisture is exhaled by frequent plowing, and no more left but to give the clay a degree of cohesion sufficient for fixing the roots. The nice point is, the time of applying the plough after much rain. To plow wet, kneads the parts together: on the other hand, the ground must not be suffered to turn hard. Between soft and hard is the proper condition for plowing; which may be known by the mouldering of the earth that is raised by the plough. During winter, clay may be stirred in a moister state, than during summer: frost prevents cohesion: heat promotes it. The management of light soil is very different. It is easily pulverized; but the difficulty

ling under ground, and are expert miners. I speak not of moles and mice, whose subterranean walks and alleys are obvious to the eye; but of worms, beetles, ants, wasps, *&c.* whose works escape observation. Their excavations may in a long tract of time render the soil extremely porous. The fine earth they dig out, is left upon the surface, and blown away by the wind.

* Part 2. chap. 1. sect. 1.

ficulty is to preferve its moifture. A long drought, by extracting much of its moifture, renders it the lefs fit for vegetation; and to plow it in that ftate in dry weather, would render it entirely unfit. The only remedy is rain; and if drought fet in, it ought to be rolled immediately after plowing.

The proper time of fowing and harrowing, is when there is no more moifture than fufficient to give the foil a proper confiftence; and I conjecture, that the fame degree of moifture is the fitteft for making the feed fpring. Reflect upon the making of malt: a certain degree of moifture is neceffary for fermentation: too much checks it. Let rolling immediately follow, to prevent as much as poffible any more evaporation. Where feed is to be fown in winter, or early in fpring, it is right to plow fometime before, in order that the fuperfluous moifture may evaporate; for at that feafon there is no fear of exhaufting the moifture. But late in the fpring, if the feafon be dry, fow the feed immediately after plowing. The plants in their courfe of growing, return to the foil, during night, part of the moifture they draw from it during day. Their roots at the fame time, fpreading in every direction, keep the foil in conftant motion, and prevent it from turning hard.

I clofe this fection with an effect of plowing, the moft important of all, becaufe it holds in all ſoils leſs or more. Plowing keeps the ſoil looſe for roots to take their natural courſe, and open for admitting air, dew, and rain. Dew in particular, which falls in plenty during ſummer, when moſt wanted, is loſt upon hard ſoil, being exhaled by the next ſun; but it ſinks deep into looſe ſoil, and is ſheltered from the ſun's power. Ground ſtirred before winter, is not only laid open to the action of the ſun, wind, and froſt, but is early ready for a ſpring crop, beans for example. It ſoon wets indeed, but it dries as ſoon. To drench in water ground left unſtirred, may require a month or two; but then, equal time is required to dry it. The more earth is pulverized, the more water it holds *; and the more parts water is divided into, the more readily it is imbibed by plants. If the ground be rendered too looſe, rolling not only makes it ſolid to ſecure the plants againſt wind, but alſo prevents evaporation.

How beneficial it is to keep ſoil open for the admiſſion of nutritive matter, will appear from the following facts. Stiff ſoil gains little by reſt; for as rain and dew get no admittance, they are ſoon carried off by evaporation. But ſoil, if tolerably open, improves by reſt. I ſuſpect that it gains little by the paſturing of cattle; for what they take away in fat with what they perſpire, will nearly

* Part 2. chap. 1. ſect. 1.

nearly balance the dung they leave : but it gains by the nutritive matter that rain and dew deposit in it : the rain may evaporate, but it leaves the nutritive matter. This case resembles salt deposited in the sea by rivers : water is evaporated from the sea, but the salt is left behind ; and hence the saltness of the sea. The nutritive matter thus left in the earth, is again diluted in rainwater ; and though not attracted by the air, is attracted by plants, and enters into the orifices of the roots along with the water in which it is dissolved. There is another cause that has a share in the improvement of an open soil, and that is air, which, with its contents, enters by attraction into the bosom of an open soil *. The operation is indeed slow, because the attraction has no effect but in contact or near it ; yet in time, the quantity of nutritive matter attracted with air, may be considerable. Du Hamel reports, that the rubbish of a mud wall made good manure, though the mud had been taken from a very poor soil. Grassy sod, used as a covering for cottages, turns good manure when it has lain long upon a house. The walls of a fold for sheep, being composed of sod, make good manure : when thrown down and mixed with the soil, they afford a better crop than the fold within, though enriched with the dung of the sheep. Among many

* Part 2. chap. 1. sect. 1.

many advantages of fallowing, the exposing to the air a new surface from time to time, is one; by that means, every part of the soil draws air with the vegetable food it contains. Columella, book 2. chap. 4. advises ground to be reduced to dust by plowing. And he quotes a saying of the ancient Romans, That that land is ill plowed which wants harrowing after the seed is sown.

2. MANURES.

THE operations of nature, hid from the ignorant and not always obvious to the learned, break out sometimes into broad day-light. Did animal bodies after death wither and dry without dissolving, this earth could not long have been a habitation for men: their utmost efforts would have been insufficient to remove dead carcases out of the way. Happily, putrefaction comes to their relief: dead bodies dissolve and mix with the soil, without leaving a trace behind. Putrefaction is a curious process of nature: air, moisture, heat, all of them, contribute; but too much, or too little, is an impediment to the process.

On the surface of this globe, a process is continually going on, unregarded by the vulgar, being too familiar to draw their attention; and yet, illustrious among the works of Providence for its beneficial effects. Plants and animals are generated,

rated, arrive at maturity; and after ſerving the purpoſes of nature, decay and rot. But the proceſs ends not there. Loathſome putrid matter, from which we avert the eye, is made ſubſervient to an excellent purpoſe, namely, renovation of plants; and the proceſs goes on without end.

Manures are of two kinds. One attracts water and is attracted by it, dung for example, ſalt, calcined limeſtome, commonly called *quicklime*, or ſimply, *lime*. Another neither attracts water nor is attracted by it, ſhell-marl for example, clay-marl, ſtone-marl, raw limeſtone beat into powder.

Of all manures, dung is the moſt univerſal. A ſoil naturally ſtiff turns free and open, in proportion to the quantity of dung beſtowed on it. Reduce clay into a dry powder: moiſten it with water, and form it into a ball: repeat the operation at pleaſure, it ſtill returns to its original hardneſs. But moiſten it once or twice with the juice of a dunghill, it becomes mellow, and never recovers its hardneſs. Dung therefore renders clay fertile, by opening it and giving admiſſion to water. It does more: it makes every ſoil retentive of water. Examine a kitchen-garden that has been often dunged in the courſe of cropping: it will be found moiſt above any neighbouring ground of the ſame original ſoil.

As dung is compoſed of putrefied vegetables or of animal excrements, it is natural to think that it contains

contains more or leſs vegetable food. This however goes not beyond a conjecture : a plant or an animal may contain abundance of vegetable food ; but we are not certain that this is the caſe after putrefaction : it may by that proceſs be converted into a different ſubſtance : ſuch converſions in natural operations, are far from being rare. But if vegetable food be contained in dung, which is the moſt likely, another uſe of it is to depoſit in the ground its vegetable food, which, being diſſolved in water, is imbibed by plants, and converted into their ſubſtance. And from an experiment mentioned above it appears, that water impregnated with dung, is of all the greateſt nouriſher of plants.

A third uſe of dung is, to promote vegetation by raiſing a kindly heat in the ground. The ſun-rays produce the ſame effect upon ground rendered black by culture ; for it is a property of all black bodies to attract and abſorb rays of the ſun *. Heat is beſt promoted by hot dung ; which therefore ſeems the moſt proper for corn. Whether hot dung be the beſt for making a ſoil retentive of moiſture, remains to be aſcertained by experiments. But I incline to think, that dung thoroughly putrefied, and conſequently cool, is in its beſt condition for graſs ; as it can be equally ſpread to give every plant its ſhare. It is alſo

* Part 2. chap. 1. ſect. 1.

alſo in its beſt condition for a kitchen-garden: green dung infects pot-herbs with an unſavoury taſte, and ſometimes with a diſagreeable ſmell.

Let a dunghill remain years without ſtirring: it is reduced in appearance to fine earth; which however has very little natural earth in it, as there is very little natural earth in vegetables, or in animals that feed on vegetables. Very few natural earths equal this vegetable earth in fertility: and it is a kind diſpenſation of Providence, not only that dung is a great fertilizer, but that when it becomes vegetable earth, it proves the beſt ſoil for vegetables. In corn-countries, the ſurface-earth comes in time to be moſtly vegetable: were it inferior to natural earth, corn-countries would long ago have been rendered barren and unfit for agriculture.

From dung I proceed to other manures. Limeſtone, ſhell-marl, clay-marl, ſtone-marl, are all of them a compoſition of calcarious earth with other ſubſtances. Sand with calcarious earth makes limeſtone *. The ſhells of fiſh are almoſt entirely calcarious; and theſe ſhells ſoftned and reduced to powder in water, are called *ſhell-marl*. Clay-marl is a compoſition of calcarious earth and clay. Stone-marl is a compoſition of clay, ſand, and calcarious earth: it is the ſand that hardens it;

* I have heard of limeſtone almoſt entirely calcarious, with little or no mixture of ſand.

it; and according to the proportion of ſand, it approaches to limeſtone or to clay-marl *.

It is obſerved above, that powdered clay is ſuſpended in water, till by the ſuperior force of gravity it fall to the bottom †. But, as far as I know, water has not the power of diſſolving any ſort of earth, calcined limeſtone alone excepted. A ſmall quantity of calcined limeſtone, a pound for example, will impregnate a vaſt quantity of water, with no loſs of bulk, and with a loſs of weight ſcarce perceptible. Calcined limeſtone thus impregnated called *lime-water*, diſcovers itſelf to the taſte though not to the eye. But this effect is confined to calcined limeſtone; for between water and calcarious earth in its natural ſtate, there appears no elective attraction: water poured on ſhell-marl comes off pure, carrying nothing along with it. Nor do clay or ſtone-marl differ, even when reduced into powder.

Vegetation is more promoted by weak lime-water than by pure water. Two beans every way equal were ſet in pots filled with earth from the ſame heap: the one was moiſtened with lime-water, the other with pure water: the firſt was by far the quickeſt grower, and the moſt vigorous. Hence one benefit of quicklime: it converts

* See Dr Ainſlie's accurate eſſay on marl. Edinburgh Eſſays phyſical and literay, vol. 3. art. 1.

† Part 2. chap. 1. ſect. 1.

verts rain into lime-water, which is a great fertilizer. This effect however is but temporary, as will thus appear. Quicklime is limeſtone deprived of its air, by the force of fire: but quicklime expoſed to the air, attracts air; and in time becomes again limeſtone as originally; conſequently unfit to make lime-water *

Quicklime may have an effect on land as well as on plants. It is highly probable, that it opens clay-ſoil to admit water that formerly reſted on the ſurface. How otherwiſe can it be explained, that liming renders clay-ſoil drier? May it not alſo have the effect to increaſe the retentive power of a looſe ſoil? This ſeems probable, if what Young the itinerant farmer ſays hold true, that lime has a much greater effect upon looſe moor than upon any other ſoil.

Salt is powerful; and an overdoze of it does more miſchief than of any other manure. It is ſoluble in water, and by that means enters the mouths of plants. Its effect then muſt be the ſame with that of lime-water; and conſidering how ſparingly it ought to be laid on land, it is not obvious what other effect it can have.

As nothing can enter the mouths of plants but what is diſſolved in air or water, calcarious earth in its natural ſtate cannot enter. Yet ſhell-marl, clay-marl, ſtone-marl, compoſed moſtly of calcarious

* Edinburgh Eſſays phyſical and literary, vol. 2. art. 8.

rious earth, contribute undoubtedly to fertility. If theſe manures cannot furniſh nouriſhment to plants directly, they muſt produce that effect indirectly, by fitting a ſoil to retain moiſture, or by preventing moiſture from acting as a cement, or by both. They certainly have the effect to keep ground from hardening: they render clay looſe and ductile, and prevent its being hardened by water. Whether they increaſe the power of any ſoil to retain water, is left to experiment.

An overdoſe of ſhell-marl, laid perhaps an inch thick, produces for a time large crops. But at laſt it renders the ſoil a *caput mortuum*, capable neither of corn nor graſs; of which there are too many inſtances in Scotland: the ſame probably would follow, from an overdoſe of clay-marl, ſtone-marl, or pounded limeſtone. How is this to be accounted for? Of one thing we are certain, that by ſuch overdoſe, light ſoil is rendered ſo looſe as to be moved by the wind; and that even clay-ſoil is rendered ſo ſoft, as to receive the impreſſion of the lighteſt foot at every ſtep. Is it not then probable, that the ſoil is rendered ſo open, as to retain little or no water? But then, how comes the land to bear any crop at all? I am reduced to another conjecture, that ordinary plowing once a-year, is not ſufficient to mix with the ſoil ſuch a quantity of manure; and that an intimate mixture requires ſeveral years. Even a moderate doſe of lime requires more than a year by by ſuch management for an intimate mixture; for which reaſon, the ſecond crop after lime is always better than the firſt, and the third frequently better than the ſecond. Now as the ſoil is opened by that part only which is mixed with it, the cropping may go on ſeveral years, before ſuch a quantity of the overdoſe, is mixed as to occaſion a total ſterility. This conjecture may be brought under the touchſtone of an experiment. Before or after harveſt, let an overdoſe of ſhell-marl be intimately mixed with the ſoil by reiterated plowings and harrowings. If the barley ſown next ſeaſon fail by lack of moiſture, the conjecture will be converted into a certainty.

Quicklime is of a nature very different from calcarious earth in its natural ſtate: by the latter, land is rendered ſo looſe by an overdoſe as to hold no water: by an overdoſe of the former, it is hardened to ſuch a degree as to be impervious to water or to the roots of plants. Several ſpots in the Carſe of Gowry, are thus rendered ſo hard as to be unfit for vegetation.

The quantity of calcarious earth in clay-marl, is frequently a half, and ſometimes more. Five hundred cart-loads of clay-marl laid on an acre, are found not to be an over-doſe. Suppoſing the half to be calcarious earth, and reckoning a cart to hold ſix bolls; here are fifteen hundred bolls of calcarious earth laid on an acre. Yet a far leſs quantity of ſhell-marl has been known to render the

the ſoil a *caput mortuum;* tho' there is not diſcovered any chemical difference, between the calcarious earth in clay-marl and that in ſhell-marl: they both equally are converted into lime by the force of fire. Ignorance of nature, betrays us at every turn to doubts and difficulties. May it not be conjectured, that calcarious earth, by entering into the compoſition of an animal body, becomes a more powerful manure than when mixed with earth? There is an argument from analogy to ſupport that conjecture. Calcarious earth changes its nature by the action of fire; and why may it not ſuffer ſome change by being made part of an animal body?

Every particular in the preſent chapter, whether relative to the plough or to manure, is perfectly agreeable to the general propoſition, That air and water with what is diſſolved in them, make the nouriſhment of plants. To cultivate land in ſuch a manner as to retain a proper quantity of air and water, is in all probability the chief or only means for making it fertile. In that view, I have all along warmly recommended pulverization; becauſe the more a ſoil is pulverized, the more water it will hold, and the more retentive it will be of it. No mode of huſbandry tends more to pulverization than horſe-hoeing; nor any inſtruments more than the brake and the harrows above deſcribed. Soils are very different with reſpect to their power of attracting and retaining air and water. Let the diligent farmer make accurate experiments for aſcertaining that difference, and for increaſing that power: no inquiries tend more to the improvement of agriculture. Our attempts to make a ſoil perpetually fertile, will probably fail; but our hopes of approaching it, may be crowned with ſucceſs.

With reſpect to the contents of this Part in general, I have to obſerve, that in natural philoſophy, of which the ſcience of agriculture is a branch, queſtions occur of two kinds. Firſt, Will a certain event happen in given circumſtances? Second, Suppoſing the event, what is the cauſe? To queſtions of the firſt kind, the anſwer is *fiat experimentum.* All that can be done with reſpect to the other kind, is from a number of analogous facts, to form a general rule or law of nature. Such rules at the ſame time ought to be admitted with caution, even after the cooleſt induction. But thoſe who are ardent for knowledge, cannot eaſily ſubmit to the ſlow progreſs of philoſophy: they are always in a hurry to draw concluſions, and hurry commonly leads them into error. Lord Bacon fancifully compares knowledge to a ladder. Upon the firſt ſtep particular truths are diſcovered by obſervation or experiment. The next ſtep is to collect theſe into more general truths; from which the aſcent is to what are ſtill more general. There are many ſteps to be taken before we arrive

rive at the top; that is, at the moſt general truths. But impatience makes us endeavour to leap at once from the loweſt ſtep to the higheſt: we tumble down, and find with regret that the work muſt begin anew.

To conclude. Here is my theory of agriculture, diſplayed at full length; which is freely ſubmitted to the public, againſt whoſe judgement there lies no appeal. But let it be kept in view, that it is ſubmitted as probable only, not as certain. It would require the life of an antediluvian, to make all the experiments that are neceſſary, for piercing to the foundation, and for reſolving all into clear principles. My life at any rate is too far advanced, for an undertaking ſo extenſive. I found an impulſe to expoſe this theory, naked as it is; and I gave way to the impulſe, becauſe I flatter myſelf, that it may afford ſome light in tracing the operations of nature. One advantage it has above ſeveral other theories, that it can be ſubjected to the touchſtone of experiments, many of which are ſuggeſted above. By ſuch experiments, ſagaciouſly conducted, it muſt ſtand or fall.

APPENDIX.

ARTICLE I.

IMPERFECTION OF SCOTCH HUSBANDRY.

A Man can never have thorough confidence in his road, till he be made acquainted with the by-paths that miſlead him; and to be made acquainted with the errors of our neighbours, is the high way to good huſbandry. My preſent purpoſe, is to delineate the imperfect ſtate of Scotch huſbandry, not only as formerly practiſed every where, but as practiſed at preſent in moſt places. To contemplate the low ſtate of their country in the moſt important of all arts, cannot fail to excite ambition to excel in the few who are ſkilful, and to rouſe imitation in others.

Our crops in general are very indifferent; and how can it be otherwiſe, conſidering our inſtruments of huſbandry, which are ſadly imperfect? What can be expected from them in a poor ſoil, when they perform ſo little even in the richeſt? Our crops accordingly correſpond to our inſtruments.

From

From many examples it is made evident, that our ſoil and climate are capable of producing draught-horſes, patient of labour, and ſingularly hardy. Yet the breed is ſo much neglected, that they are commonly miſerable creatures, without ſtrength or mettle. Did landlords attend to their intereſt, they would be diligent to improve the breed. Why do they not reflect, that the ſame farm-ſervants with better horſes, would double the ordinary work? By improving the breed, they would draw more rent from their tenants, without lying any additional burden upon them. With reſpect to oxen, there is no care taken either in the breeding or feeding. How eaſy is it for a gentleman to procure a good bull for his tenants? and from the little care of providing food for draught-oxen, one would ſuſpect it to be a general opinion, that they require no food. In ſummer they are turned out into bare paſture, ſcarce ſufficient for ſheep. In winter, a ſmall bottle of ſtraw, not above a ſtone weight, is all that is allowed them in the twenty-four hours.; which after the turn of the year, being dry and ſapleſs, affords very little nouriſhment. What can animals ſo fed do in a plough? And yet ſuch is the ſtupidity of many farmers, that inſtead of adding to the food, they add to the number; as if it would mend the matter, to add cattle that can ſcarce ſupport their own weight. One unaccuſtomed to ſee ten oxen in a plough led on by two horſes,

horſes, cannot avoid ſmiling. With his goad the driver beats the horſes, and pricks every ox as he advances. He then runs forward twenty yards to beat the horſes a ſecond time and prick the oxen. Some of the oxen in the mean time, inſtead of drawing, are found hanging on the yoke, and keeping others back. It is indeed next to impracticable, to make ten weak oxen in a plough, draw all at the ſame time. Nor is this the only inconvenience. A great number of oxen by ſuch management, are requiſite for ſtocking a farm; the expence of which is not always within the reach of the moſt induſtrious. In a year of ſcarcity beſide, the beaſts are actually ſtarved. And what is worſt of all, the tenant, in order to get ſtraw for his cattle, is commonly neceſſitated to threſh out his corn, without waiting for a market, or having a granery for it.

Our farmers, led entirely by cuſtom, not by reflection, ſeldom think of proportioning the number of their working cattle, to the uſes they have for them. Hence, in different counties, from ſix to twelve oxen in a plough, without any regard to the ſoil. Seldom it is, that more than four good beaſts can be neceſſary, if the proper time for plowing be watched.

The diviſion of a farm into infield and outfield, is execrable huſbandry. Formerly, war employed the bulk of our people: the remainder were far from ſufficiently numerous for cultivating even that ſmall proportion of our land which is capable of the plough. Hence extenſive farms, a ſmall part of which next the dwelling, termed *infield*, was cultivated for corn: the remainder, termed *outfield*, was abandoned to the cattle, in appearance for paſture, but in reality for ſtarving. The ſame mode continues to this day, without many exceptions, though neceſſity cannot be pleaded for it. But cuſtom is the ruling principle that governs all. Sad is the condition of the labouring cattle; which are often reduced to thiſtles, and withered ſtraw. A ſingle acre of red clover would give more food than a whole outfield; yet how common is the complaint of tenants, that they are diſabled from carrying on any ſummer-work, for want of food to their horſes; a ſhameful complaint, conſidering how eaſy the remedy is.

Cuſtom is no where more prevalent than in the form of ridges. No leſs high than broad, they are enormous maſſes of accumulated earth, that admit not croſs-plowing, nor any plowing but gathering and cleaving. Cuſtom and imitation are ſo powerful, as that our ridges are no leſs high in the ſteepeſt bank, than in the flatteſt field. Balks between ridges are equally frequent, though invincible obſtructions to good culture. It would puzzle one at firſt view to explain, why ſuch ſtrips of land are left untilled. They muſt have been reſerved originally, as a receptacle for ſtones, thrown

thrown off the tilled land ; and huſbandmen were led by imitation to leave ſuch ſtrips, even where there were few or no ſtones.

The proper time for plowing or harrowing, is when the ſoil upon ſtirring moulders into ſmall parts. This is not obſerved by farmers ſo carefully as it ought to be. How common is it to ſee even a clay ſoil plowed, when ſoaked in water, or when hard like a ſtone. Little attention is given to what may be termed *the froſt-preparation ;* which is, to open the ground before winter, in order that froſt may pierce deep, and mellow the ſoil.

Shallow plowing is univerſal, without the leaſt regard to deepneſs of ſoil. The temperance of our people may be a proper ſubject for ironical praiſe ; for though nature affords commonly ten or twelve inches of ſoil, they are humbly ſatisfied with a half or third.

Ribbing is a general practice, though the ſlighteſt reflection is ſufficient to make it evident, that to leave half of the land untilled, muſt be wretched huſbandry.

Summer-follow has of late years crept in, and is now common in three or four counties. In the reſt of Scotland, for want of ſummer-fallow, there is a continual ſtruggle for ſuperiority, between corn and weeds. Do not ſuch provoking farmers ſee, that it is fruitleſs to manure land over-run with weeds? Do they not obſerve, that the

the manure they beſtow encourages weeds as much as corn; or rather, that it envigorates the weeds to deſtroy the corn? Make a progreſs through Scotland, you ſee ſtubborn weeds in every corner ſcattering their ſeed, and fouling the ground more and more. It is an eaſy work to cut down weeds before they go to ſeed. Would not one think, that work ſo eaſy would never be neglected? and yet it is never done. A Scotch farmer behaves worſe than Eſau: the latter got a meſs of pottage for his birthright; the former ſurrenders his to weeds, without any recompenſe.

There is ſcarce ſuch a thing practiſed, as to harrow before ſowing. The ſeed is thrown into rough uneven ground, and the half is buried.

The roller is a moſt uſeful inſtrument. It was unknown till lately; and even at preſent is very little uſed.

With regard to rotation of crops, a moſt important article, there is great ignorance among our farmers. As it would be tedious to enter into particulars, I refer to ch. 7. where that ſubject is treated of.

Our farmers ſhow very little ſkill in harveſt-work. I confine myſelf to a ſingle inſtance. The ſheaves are bound up with a rope, compoſed of two lengths of the corn, twiſted together; which makes the ſheaves commonly of a monſtrous ſize. The binder, preſſing hard with his knee, binds the

the ſheaf ſo cloſe, as with difficulty to admit his finger. The weather muſt be extremely favourable, if it be ſufficiently dry in a fortnight, to be ventured in a ſtack; it commonly muſt ſtand in the field three weeks. Let any one conſider the riſk of the crop in various weather, ſuch as happens ordinarily in autumn. Nor is this all. Such ſheaves are not only unhandy, but are apt to looſen in being carried to the ſtack, or from the ſtack to the barn. A ſheaf ſhould never exceed what can be ſlightly bound together with a ſingle length of the corn: it is fitter to be ſtacked in a week, than ordinary ſheaves in three.

No branch of huſbandry is leſs underſtood, than manure. A dunghill is a very improper bed for corn: lime and marl are ſtill more improper; for nothing will grow on them. Hence it is apparent, that the intimate mixture of manure with the ſoil, is the great circumſtance for vegetation. In order to that end, the ſoil ought to be highly pulverized, and the manure divided into its ſmalleſt parts. With reſpect to dung in particular, it ought to be carefully mixed in the dunghill, not neglecting to divide by the hand any lumps that may be in it. Let our farmers ſay, whether they are ſo accurate. Nothing more ordinary, than heaps of dung withering in the field, incapable to be intimately mixed with any ſoil. Nothing more ordinary, than dung laid on the dunghill in barrowfuls,

barrowfuls, without being ſpread or mixed with what was there before.

A potato is a moſt uſeful plant, and, when properly cultivated, affords a plentiful crop. It is a great reſource to the labouring poor, being a nouriſhing food that requires very little cooking. We have been afflicted of late years with very bad ſeaſons, which, but for that reſource, muſt have driven many of our people from their native country. Yet potatoes to this day continue to be propagated in lazy-beds. Expert farmers, not many in number, raiſe them with the plough at the twentieth part of the expence. This method has beſide two other advantages: it leaves the ſoil in the beſt ſtate for ſubſequent crops; and the potatoes are more palatable than what are raiſed in a lazy-bed.

Swine make a profitable article of huſbandry, very little attended to in Scotland. They are fed at a ſmall expence, and yet make moſt nouriſhing food. Every perſon who has a cow, ought alſo to have a pig. This is univerſal in England: it is creeping into Berwickſhire, but in few other places as far as I know.

Very few farms in Scotland are juſtly proportioned: ſome are too ſmall; the bulk of them too large. The former lead to a habit of idleneſs; the latter into a habit of ſlovenlineſs, by want of power to do juſtice to every part. There is not an article in huſbandry more eſſential, than

to

to adjuſt a farm to the ſkill and ability of the tenant.

ARTICLE II.

A BOARD FOR IMPROVING AGRICULTURE.

IT is a maxim in politics, that every country will be populous in proportion to the fertility of its ſoil; upon which account, agriculture is the moſt uſeful of all arts. And yet it is a ſad truth, that in Scotland, this art has advanced not far beyond the firſt ſtage of its progreſs. In England indeed, it has made a much greater advance; and yet far inferior in perfection to Engliſh manufactures and commerce. Agriculture is carried on every where without a ſchool; and for that reaſon, is commonly thought to require no ſchooling. Can a Britiſh miniſter embrace any meaſure more patriotic, than to encourage agriculture and its profeſſors? No other meaſure would ſo effectually aggrandize Britain. A ſmall ſhare of the money and attention beſtowed on raiſing colonies in America, would have done wonders at home. And yet, mark the ſtriking difference: our arts are our own, which we never can be deprived of while induſtry remains: in the very conſtitution of our colonies, on the contrary, there are cauſes of ſeparation, that grow daily

daily more and more efficacious; a wide-extended country, a fertile ſoil, navigable rivers, and a growing population. I diſregard the preſent rebellion of our Americans: for they will ſoon be reduced to obedience. But as they derive from Britain high notions of liberty and independence, and as they are daily encreaſing in power and opulence, the æra of their total ſeparation, cannot be at a great diſtance. It. is indeed abſurd to think, that a great nation, in the vigour of proſperity and patriotiſm, can be kept in ſubjection by a nation not more powerful, enervated by luxury and avarice. Let us not however diſpond; for if agriculture be carried on but to the perfection that our ſoil and climate readily admit, it will amply compenſate the loſs of theſe colonies.

Books are uſeful for advancing huſbandry, otherwiſe this little treatiſe ſhould not have ſeen the light. But books are far inferior to living inſtructors, who convey knowledge by practice as well as precept. We have a board for manufactures and fiſheries, a wiſe inſtitution which has been attended with great ſucceſs. Why not alſo a board for agriculture? Is agriculture a leſs uſeful art than theſe mentioned? or does it leſs require inſtruction? Hartlib, in his legacy, laments that no public director of huſbandry had ever been eſtabliſhed in England. The preſent time is in Scotland the happieſt for ſuch an eſtabliſhment.

Before

Before the union of the two kingdoms, our people were ſo benummed with oppreſſion, that the moſt able director would have made no impreſſion. Freedom has braced their nerves, and has made them take heart to be induſtrious. They liſten to inſtruction: let them perceive their intereſt, and they will cheerfully practiſe what they are taught. A board for agriculture would among us have wonderful ſucceſs in many important articles. Conſidering the quantity of waſte land even in our beſt-cultivated counties, it is not too ſanguine to hope, that our corn-crops may be doubled. What a bleſſing would this be to Scotland, which for many years has been reduced to import great quantities? Our horſes and horned cattle, are far inferior to what may be produced by good management. Our ſheep weigh not above ten pounds a quarter, nor their wool above two pounds. The ſoil by good culture would feed ſheep to the weight of twenty-four pounds a quarter; carrying from ſix to ten pounds of wool, a valuable acquiſition to the woolen manufacture. Lambs in ſeveral inſtances have been advanced to twelve ſhillings a-head, and wedders to forty ſhillings. Theſe are but a ſpecimen of the various improvements that might be perfected by ſuch a board.

The plan I have in view, is ſimple. Let the board conſiſt of nine members, the moſt noted for ſkill in huſbandry, and for patriotiſm. As I propofe no reward to thefe gentlemen but the honour of ferving their country, the choice will not be difficult: in lucrative employments, perfonal connections have more influence than perfonal merit; and it is avarice only that fets people at variance. Where perfonal merit is the fole object of choice, there is feldom much difference in opinion. And to have a right fet of members at firft is of the utmoft importance. If deficient in knowledge, they will have no influence, and perhaps be fneered at. But let men be chofen who have the public voice for them: they will have great authority, and every direction of theirs will be obeyed. To eafe the board in the laborious branch of their bufinefs, they ought to be provided with an able fecretary, to minute their proceedings, to write their difpatches, and to carry on their correfpondence, foreign and domeftic. As punctual attendance is neceffary, the good behaviour of fuch an officer may well intitle him to a falary of L. 100 yearly; with the addition of L. 30 more in a year of extraordinary bufinefs, at the difcretion of the board; but not unlefs all the members be unanimous. A larger falary would be an object of intereft, and foon degenerate into a finecure.

A regular meeting once a month may be fufficient; with liberty to thofe who have moft leifure, to meet at intervals for expediting what may require difpatch. It would cramp the proceedings

ceedings of ſuch a board, to confine it to a quorum. As there cannot be any ſelf-intereſt to create a bias, thoſe who meet ought to have the power of the whole; and what they tranſact ought to be final, if not altered by a greater number the next monthly meeting.

The things neceſſary to be undertaken by this board at the commencement of their operations, will require much labour and ſagacity. The firſt is, to make out a ſtate of the huſbandry practiſed in the different counties; in which notice muſt be taken of the climate, of the ſoil, of the mode of cropping, and of the inſtruments of huſbandry, noting the prices of all the particulars that enter into farming. The next is written inſtructions for improving huſbandry, ſuited to the ſoil and ſituation of every diſtrict; with ſpecial reference to the preſent practice, ſhowing where it is defective or erroneous, and propoſing the cheapeſt and moſt effectual corrections. Theſe preliminaries being ſettled, the ordinary buſineſs of the board, may be carried on eaſily and commodiouſly. In the firſt place, there is a neceſſity for an inſpector, named by the board, to make a progreſs from time to time in ſucceſſive places, for reporting the progreſs of the improvements directed, and for giving inſtruction in caſes that cannot ſo clearly be put in writing. In this progreſs, ſpecial notice ought to be taken of the beſt-conducted farms, whether by landlords or tenants.

A

A few ſilver medals beſtowed on the moſt deſerving, will rouſe emulation in all, and promote induſtry. Second, this board will conſider it as a capital branch of buſineſs, to anſwer queries, and to ſolicit a correſpondence with men of ſkill. Third, they ought carefully to inform themſelves of every invention that tends to improve the art, and to publiſh what they think uſeful. Fourth, premiums ought to be propoſed, and diſtributed among thoſe who profit the moſt by the inſtructions of the board. Theſe premiums ought to be ploughs, harrows, carts, conſtructed after the beſt models; which beſide exciting induſtry, will be a means to introduce the beſt huſbandry-inſtruments. Fifth, in no other reſpect would a board of agriculture be ſo uſeful, as in directing proper experiments. Agriculture, though the prime of arts, is far from perfection in any country. This in part is owing to its complex nature; but chiefly, to the length of time that is neceſſary to aſcertain, by a courſe of experiments, any capital point in theory or practice. The life of man is too ſhort for ſuch an undertaking. The only remedy is to employ many hands upon different experiments; which cannot be done effectually, but under the direction of a board that never dies. Let liſts be made from time to time, of the points that are capable to be aſcertained by experiments: let proper experiments be ſuggeſted: let theſe experiments be diſtributed

diſtributed among perſons of ſkill. And when their ſucceſs is reported, the concluſions that may be drawn from them, ought to be publiſhed. This would be the moſt effectual method that ever has been contrived, to ripen knowledge in huſbandry. To enliven this branch of buſineſs, premiums ought to be propoſed, lucrative as well as honorary.

Of the premiums to be diſtributed, ſcarce any would be of more general benefit than to the beſt hand-hoers under the age of fifteen. Boys in driving the cart or the plough find exerciſe for their limbs; but in huſbandry the arms are ſeldom exerciſed till they be full grown. I reliſh hand-hoeing for keeping ground clean: I reliſh it more for the opportunity it gives to exerciſe the arms of young creatures, male and female, from ten upward: give them only hoes of different ſizes adapted to their ſtrength. I venture to affirm, that the ſtrength of a man's arms who has been employ'd in hand-hoeing from his tender years will be far greater, perhaps a third, than if they had never been exerciſed till he was fully grown. This would be a great advantage in ſeveral employments, civil and military, as well as in agriculture. Add another advantage. People accuſtomed from their tender years to keep ground clean, will contract an early averſion to weeds, and declare perpetual war againſt them. My labourers have good kitchen gardens, where

where onions, leeks, cabbage, turnip, and potatoes are ſown in drills. The hoe is conſtantly employed by their wives or their children. You may ſee a dirty face among them, but not a dirty garden.

To make the board proceed with ſpirit, a book or pamphlet ought to be publiſhed annually, containing their tranſactions during the preceding year. The profit of the work, is a perquiſite to the ſecretary; which will encourage him to beſtow his utmoſt ſkill in the compilation.

To procure public favour, men of character and knowledge may be introduced by the members at their monthly meetings, to aſſiſt in their deliberations.

The choice of proper members, is the capital point: the whole depends on it. The choice is the more difficult, as it muſt be confined to gentlemen who reſide in Edinburgh, ſome part of the year; becauſe from others punctual attendance cannot be expected. It would be unſafe to leave the choice to members of parliament; who even againſt their private ſentiments, are obliged to ſolicit for their friends and voters, without regard to merit. The choice muſt not be left abſolutely to the chief miniſter: who, at ſuch a diſtance, is ſeldom perſonally acquainted with the beſt qualified. The ſafeſt method I can think of is, that the juſtices of peace of each corn-county, ſhould at a quarter-ſeſſions name one. Out of

theſe,

theſe, the nine members are choſen by the crown.

In Scotland, many noblemen and gentlemen, ſkilful in huſbandry and zealous to promote it, would make excellent members but for their diſtance from the capital. To require conſtant attendance from ſuch would be too great a burden; but to intitle them to act when they ſhould find it convenient, under the title of honorary members, would add great luſtre to the board.

The choice of a member to ſupply a vacancy, is a matter no leſs delicate. A ſociety of gentlemen, who ſerve for honour not for profit, are well intitled to chooſe their companion. But to avoid faction, which would be ruinous in ſuch a ſociety, the choice ought to be unanimous. The diſſent of a ſingle member need not be regarded; but if two diſſent, the choice muſt be in the crown. If a member be abſent three ſucceſſive monthly meetings, without an excuſe approved by the board, he is to be held as having deſerted his office, to make way for the election of a new member.

The election of a ſecretary is a point ſtill more delicate. The board ought naturally to have the choice of their own ſecretary; but in caſe of a diviſion, the diſſent of three from the other ſix, ſhall transfer the election to the crown.

The royal ſociety in London, is perhaps the only ſociety in the world, that has flouriſhed ſo long, with

with no other motive but thirſt for knowledge. The members have now an additional motive, which is the reputation of being enliſted in a ſociety, ſo illuſtrious. In the preſent low ſtate of patriotiſm, affection to one's country is not alone ſufficient, to preſerve long in vigour a board of agriculture. Luckily, there is an additional motive, inherent in the very nature of the inſtitution. Money is neceſſary to carry on the operations of the ſociety; and the diſtribution of that money among perſons of merit, will be a conſtant entertainment to the members. A great ſum would be a temptation to miſapply it. Therefore, no moreought to be put in their power, but what is barely ſufficient to carry on the management with ſucceſs. Beſide the ſecretary's ſalary, L. 500 yearly diſcreetly diſtributed may be ſufficient. And I boldly affirm, that ſuch a ſum cannot be laid out with more advantage, whether the public be regarded, or the good of a valuable portion of our people.

The houſe poſſeſſed by the truſtees for manufactures, will afford good accommodation to both ſocieties; and ſeveral of the acting truſtees are qualified to make a figure in both.

Zeal for the proſperity of Britain, makes me ardently wiſh to have this plan extended to England. The Engliſh enjoy the reputation of being excellent farmers; and ſo they are, compared with their neighbours in France, Italy, and Spain. They

They are however far, very far, from the perfection of the art. A board for agriculture is indeed lefs neceffary in England, than in Scotland; but that England would be greatly profited by fuch an inftitution, will be acknowledged by every one who is acquainted with Englifh agriculture. I appeal to Mr Young for the following facts, extracted from his different tours; which, at the fame time, are but a fpecimen of much wrong practice mentioned by him.

Seldom is a plough feen in England with fewer than four horfes, nor is it always confined to that number; and yet, feldom are more than two horfes neceffary, if the plough be well conftructed. A great fum is thus expended upon fuperfluous horfes, which wounds the public by unneceffary confumption, is hurtful to landlords by leffening their rent, and retards the progrefs of hufbandry. Among numberlefs inftances, I mention the Ifle of Thanet, where the foil is a light loam on a chalky bottom; and yet with four horfes in each plough, they feldom pierce deeper than three inches, which is fcratching inftead of plowing. In Licefterfhire the common practice is to ufe from four to feven horfes in a plough, even where the foil is a fandy loam. With this plough they feldom do more than half an acre in a day; and yet there are gentlemen there who with two horfes plough with eafe a whole acre.

 The

The number of draught cattle is ſeldom proportioned, with any accuracy, to the extent of the farm. Frequently, no fewer than eight horſes and as many oxen, are employed in a farm of a hundred acres. With ſuch an expence, the land muſt be fertile indeed, if it afford any rent to the landlord. In ſome farms not exceeding fifty acres, ſix horſes are kept. The uſing oxen inſtead of horſes, and employing no more of them than neceſſary, would be a ſaving to England of ſeveral millions yearly. Were that improvement accompanied with a proper regulation for the poor, England would be in a higher ſtate of proſperity, than is enjoyed by any other nation.

A ſkilful rotation of crops, is far from being common. Inſtances are frequent in every part of England of the following rotations, fallow, wheat, oats, wheat. Alſo, fallow, wheat, oats, oats. Alſo, fallow, wheat, oats, barley. Alſo, barley, oats, oats. Alſo, turnip, barley, oats, oats. Even the beſt ſoils muſt be exhauſted in time, by ſuch oppreſſive cropping.

The great advantages of horſe-hoeing, are a crop, and at the ſame time a ſubſtantial fallow. And yet, horſe-hoeing, though invented in England, is not practiſed there. Many farmers do not even hand-hoe their turnip crop; and many neglect to hand-hoe their bean-crop, after being ſowed in drills.

The

The proper management of artificial graſſes, is far from being common. Of all graſſes, red clover is the moſt beneficial; and yet there are farmers, not a few in number, who baniſh red clover, as hurtful by foſtering weeds. It has indeed that effect, if allowed to grow three or four years; but why not change it every year? It is not unfrequent to ſee a field left to be covered with natural graſs. By this ſlovenly practice, the crops are not only ſcanty, but of a bad kind. In Derbyſhire particularly, a field, after three ſucceſſive crops of oats, is abandoned to nature. Worſe huſbandry is not to be met with, among the moſt ignorant farmers in Scotland.

Draining indeed is common, but conducted with little ſkill. There is no ſuch thing known in England as drains on the ſurface made with the plough; though ſuch drains poſſeſs the advantages of being cheap, effectual, and perpetual.

Let me add to theſe the following of my own obſervation. Travelling from Burrowbridge to Ferrybridge, Doncaſter, Worſop, Mansfield, Nottingham, *&c.* the land is moſtly of a ſandy ſoil. The far greater part is laid out in graſs encloſures, which give no proper return as the graſs ſoon withers in ſummer. It ought to be cropped with turnip, potatoes, barley, hay, and plenty of red clover for ſummer feeding. Such cropping would afford four times its preſent rent, beſide promoting population and the public revenue. But without

out a board of agriculture this reform cannot be made. A board of agriculture would select a few of the most promising tenants to execute their improving plans, and join with the landlords in premiums to the most deserving. Success and prosperity would prevail with others to follow their example. From Birmingham to Liverpool thro' Woolverhampton, Stafford, *&c.* mostly a sandy soil, yet no turnip and little red clover. The making hay is not generally well conducted in England. In the year 1778, the weather was both dry and hot during the time of this operation; and yet I frequently saw the hay spread on the ground lying withering whole days together. In the county of Chester, cheese is the chief product; and yet appears not to be managed to the greatest advantage. The grass enclosures are far from being rich. It is said indeed that rich grass would make the curd ferment and swell, which would occasion the cheese to be full of holes. However this be, I am certain that an acre of red clover would feed more than six of their grass acres; and if the cheese produced be richer, it may not indeed be Cheshire cheese, but it will give a better price. In a dry summer beside, their pasture grounds become early bare; and they supply the want of grass with hay. Would not green clover be a comfortable resource in such a case? But what I chiefly insist on as unexcusable, is that their cows are a heavy burden on them all winter, being

ing fed with hay and at times with ſheaves of unthreſhed corn. The ſoil is every where well adapted for turnip; which during winter would produce milk, ſufficient in butter to pay the expence of the turnip; beſide preſerving the cows in good plight for calving. It would be eaſy for a board of agriculture to ſet on foot this improvement; and as the preſent practice is of a long ſtanding, it will never be thought of otherwiſe.

From the beginning of time every ſubſtantial improvement has been ſet on foot by the landlord, who has the capital intereſt. The management of eſtates in England is generally not in the hands of the proprietor, but of his ſteward, whoſe advantage it is to ſqueeze the tenants for his own profit, and not to improve the land for that of his maſter. It is his intereſt to keep the tenants low and at his mercy; for an opulent tenant might ſtand in oppoſition and proclaim dangerous truths. When this is the caſe, is it a wonder to ſee much bad huſbandry in England?

The foregoing errors and imperfections, with an endleſs number more, would be remedied by a board, eminent for patriotiſm and for ſkill in agriculture; and farmers would fairly be directed to the road that leads to the perfection of their art. Population and induſtry would be the conſequences, with a great increaſe in the public revenue. England would become ſo proſperous and powerful, as to ſuffer little diſtreſs

from

from the loſs of its American colonies, ſhould that ungrateful people ſucceed ultimately in a total defection.

ARTICLE III.

GENERAL HEADS OF A LEASE FOR A CORN-FARM.

IN a leaſe of this kind, what chiefly ought to be in view, is to reſtrain the tenant from impoveriſhing the land, and yet leave him at liberty to improve it; reſembling a Britiſh monarch, who has unbounded power to do good, none to do miſchief. In this variable climate, the tenant muſt not be tied down to invariable rules of cropping: an unuſual ſeaſon, hot, cold, dry, or wet, will neceſſitate him, for a year at leaſt, to abandon the beſt plan of cropping that can be contrived before hand.

This obſervation is not intended to baniſh rules altogether. Some tenants, like ſome kings, may be truſted with unlimited powers. But ſuch powers would be no leſs deſtructive to the generality of tenants themſelves, than to their landlords. Tenants, therefore, like kings, muſt be fettered; but in what manner, is a queſtion no leſs difficult than uſeful. They ought not to be ſo ſo fettered as to bar improvements; nor left at liberty to do miſchief.

Before entering into particulars, it muſt be obſerved, that different ſituations with regard to manure, ſoil, and climate, require different modes of huſbandry. All that can be done in an attempt like the preſent, is to ſuggeſt a few general rules, for a landlord to chuſe upon in granting leaſes. It is his buſineſs to judge which of theſe rules will beſt fit his ſituation.

The firſt reſpects the time of endurance, which, though an important article, is unneceſſary to be enlarged on here. It is believed to be now the univerſal opinion, that without a long leaſe, it is vain to hope for an improving tenant. The moſt approved time of endurance, as the likelieſt to prevent waſte, is to fix a time certain, ſuppoſe nineteen or two nineteen years; and to add the life of the tenant who is in poſſeſſion at the expiry of the time certain. A man never loſes hope of living longer; and he will never run out ground, that he hopes yet to be long in poſſeſſion of. By this means, the tenant is deluded into a courſe of management, equally profitable to himſelf and to his landlord. But what, if, after liming or other expenſive manure, the tenant happen to die ſuddenly before reaping any profit? With a view to that event, let there be a clauſe in the leaſe, for paying to his repreſentatives

tives what ſum the tenant's profit has fallen ſhort of the expence.

Second, Aſſignees and ſubtenants ought to be excluded. For where a tenant has it in his power to make his leaſe a ſubject of commerce, he will be ſparing in laying out money on improvements.

Third, Whether the rent ought to be paid in corn or money, depends on circumſtances. Corn-rent cramps the tenant in his management; for it obliges him to ſow yearly corn of the ſame kind with what he pays. Money-rent, on the contrary, promotes good culture, in order to produce the weightieſt grain, the benefit of which accrues entirely to the tenant. There is an additional reaſon for money-rent, that the tenant, by prudence and patience, can draw a better price for his corns at the home-market, than his landlord can. The rent therefore ought to be paid in money, unleſs where there is a ſuperfluity of corn for exportation; which can be managed with more advantage by the landlord, who has all his farm-corns to export, than by the tenant who has but a ſmall quantity.

Fourth, In this country, the profit of graſs is to this day not underſtood, but by a few. Corn is the object of the generality; and that wrong bias ought to be rectified, by a clauſe confining the tenant to a certain proportion of his land in corn, a third, for example, or a half. There cannot

cannot be a general rule; becauſe it varies with the nature of the ſoil, and ſtill more with the opportunity of manure. But to give room for extraordinary improvements, an addition to the proportion of corn may be indulged, upon condition of paying   ſhillings additional rent for every acre above the proportion originally agreed on.

Fifth, A clauſe prohibiting white corn-crops to be taken in immediate ſucceſſion, will be an effectual bar againſt impoveriſhing the land. Peaſe, beans, turnip, cabbage, and potatoes are profitable crops; and red clover may be more profitable than any of them, by feeding all the farm-cattle upon it, which will ſave many acres of paſture. This and the foregoing rule, without any other precaution, will in all events ſuffice to keep the ground in good heart.

Sixth, The following, or ſome ſuch clauſe, will excite a tenant's higheſt induſtry to improve his farm, ſuppoſing it to be only for nineteen years. At expiry of the leaſe, the tenant ſhall be intitled to a ſecond nineteen years, upon paying a fifth part more of rent; unleſs the landlord give him ten years purchaſe of that fifth part. The rent, for example, is L. 100. The tenant offers L. 120. He is intitled to continue his poſſeſſion a ſecond nineteen years at the advanced rent, unleſs the landlord pay him L. 200. If he offer a ſtill higher rent, the landlord cannot turn him out, unleſs he pay him ten years purchaſe of that offer.

Seventh, As both landlord and tenant are concerned in preſerving the fences, both ought to concur in the expence. Therefore, let the care of the fences be truſted to the landlord's hedger; and whatever work is beſtowed on the tenant's fences, ſhall be paid to the hedger at ſo much *per* day. Where the preſervation of the fences is left entirely to the tenant he turns careleſs and does things by halves; where it is left entirely to the landlord, the tenant takes no care to keep his cattle from treſpaſſing.

Eighth, In order to preſerve to the landlord a privilege to plant trees, which is commonly neglected in leaſes, I propoſe that out of the leaſe be excepted certain ſpots, proper to be planted, for ſhelter, for beauty, or as not being arable; the landlord to encloſe and plant, the tenant to carry the ſtones that are neceſſary for encloſing. To encourage him to preſerve the trees, he is to have the whole weedings for the purpoſes of his farm. There may beſide be added a clauſe, encouraging the tenant to plant trees, by permitting him to cut them down for his own uſe. And the landlord is to have his choice, either to pay for what are left at the tenant's removal, or to allow him to diſpoſe of them.

Ninth, In a tenant two things are required; firſt, ſkill and induſtry for managing the farm; and, next, money for ſtocking it ſufficiently, without which, ſkill and induſtry avail not. With

reſpect

respect to both, our common law errs grossly. As to the first, a farm can never be prudently managed by a plurality; for there it holds, so many men so many minds; and yet, by law heirs-portioners succeed in a lease, as well as in other heritable subjects. To remedy the common law in leases that go to heirs, let it be provided, that the eldest heir-female shall succeed without division; or that the landlord shall have it in his power to chuse any of the heirs-female he pleases.

With respect to the other, our common law is altogether unjustifiable, as it gives the whole stocking to the other children, leaving the bare lease to the heir, without means to stock the farm anew, unless other heritable funds be left beside the lease, which seldom is the case. This is cruelly unjust, both to the heir and to the landlord. The heir has not even the benefit of collation, because it would bring a plurality of conjunct lessees upon the landlord. The heir therefore is in effect totally disinherited; as a bare lease is of no significancy without money or credit. The injustice with regard to the landlord is no less flagrant, who has thus a tenant imposed on him, from whom no rent can be expected. To preserve the lease and stocking united, which must be done by paction since law is defective, let a sum be specified in the lease, such as may be sufficient for stocking the farm; which sum the heir shall be intitled to demand from his predecessor's representatives,

representatives, unless the farm be left to him with a stocking equal in value. And the clause may be conceived in some manner like what follows. "And considering that if the said A. B. "die during the currency of this lease, his whole "moveables, not excepting the stocking of his "farm, will fall by law to his other children, by "which it may happen, that nothing is left to the "heir but the naked lease, without a stocking "or money to purchase it; therefore, to prevent "this hardship, equally prejudicial to the heir "and to his landlord, it is expressly covenanted, "notwithstanding the time of endurance above "specified, That this lease shall fall and be extinct "by the said A. B.'s death, unless he make good "to his heir effects heritable or moveable to the "extent of Sterling, the parties being sensible, that a stocking proper for this farm cannot be of value less than the said sum."

Tenth, To render the removing of tenants at the expiry of the lease more easy and certain than it is by our law, and without expence to either party, I propose the following article. Supposing a lease for nineteen years to be agreed on at a rent of L. 50, let one, two, or three years be added, binding the tenant to pay an additional rent for these years, a half more for example, or double. But with a proviso, that the tenant shall be at liberty to remove at the end of the nineteen

nineteen years, upon notifying to his landlord, three months before, his intention to remove.

ARTICLE IV.

PLANTS AND ANIMALS COMPARED.

ANIMALS are provided with various powers corresponding to their destination; some for supporting the animal frame, some for gratifying desire. Plants, in all appearance, have no feeling of pleasure nor of pain; and consequently no desires. But they are endowed with powers for preserving vegetable life, as animals are for preserving animal life. Doth not the springing of the seed, the motion of the sap, the production of leaves, flowers, fruit, *&c.* proceed from a power in plants; as the beating of the heart, the circulation of the blood, *&c.* proceed from a power in animals? There is not an argument for the latter that does not equally conclude for the former.

Next, as to the power of loco-motion. That power is more perfect in animals; but plants possess a share of it, such as is necessary for their well-being: they grow both upward and downward; and in their progress to maturity, they are continually occupying new parts of space. Plants, it is true, cannot, like animals, go out of harms

harms way; but it is curious to obſerve, how they exert that ſhare of loco-motion they are endowed with, to avoid harm. Upon the ſlighteſt touch, the ſenſitive plant ſhrinks back and folds its leaves; ſimilar to a ſnail, which on the ſlighteſt touch retires within its ſhell. A new ſpecies of the ſenſitive plant has been lately diſcovered. If a fly perch upon one of its flower leaves, it cloſes inſtantly, and cruſhes the inſect to death. The nettle never fails to ſting the hand that touches it. There is not an article of botany more admirable than a contrivance viſible in many plants, to take advantage of good weather, and to protect themſelves againſt bad. They open and cloſe their flowers and leaves, in different circumſtances: ſome cloſe before ſunſet, ſome after: ſome open to receive rain, ſome cloſe to avoid it. The petals of many flowers expand in the ſun; but contract at night, or on the approach of rain. After the ſeeds are fecundated, the petals no longer contract. The common goatſbeard cloſes up its flowers while the ſun paſſes the meridian. The pimpernel expands its leaves at ſunſet, and cloſes them at ſunriſing. All the trefoils may ſerve as a barometer to the huſbandman: they always contract their leaves on an impending ſtorm. Some plants follow the ſun, ſome turn from it. Moſt diſcous flowers follow the ſun; which has been long obſerved of the ſun-flower, while young and tender. The leaves of the mallow

low tribe follow daily the courſe of the ſun, from eaſt by ſouth to weſt. Many plants on the ſun's receſs vary the poſition of their leaves; which is ſtiled the *ſleep of plants*. Every botaniſt, after Pliny, has obſerved this in a field of clover. A ſingular plant was lately diſcovered in Bengal. Its leaves are in continual motion all day long; but when night approaches, they fall from an erect poſture down to reſt *.

A

* This curious property ſtiled the *ſleep of plants* deſerves further illuſtration by an induction of particulars. Yellow goatſbeard flowers in June. It expands its flowers about three or four in the morning, and cloſes them about nine or ten forenoon. The flowers of ſmooth ſuccory hawkweed are expanded from four in the morning till noon. The African ſowthiſtle with a poppy leaf, expands its flowers between four and ſix in the morning, and cloſes them about three hours after. The flowers of the day-lilly, expand about five in the morning, and cloſe about ſeven or eight in the evening. Wild-poppy with a naked ſtalk and a yellow ſweet-ſmelling flower, expands its flowers at five in the morning, and cloſes them at ſeven in the evening. Bindweed a little blue convolvolus, expands its flowers between five and ſix in the morning, and cloſes them in the afternoon. Roſe-coloured goatſbeard expands its flowers between five and ſix in the morning, and cloſes them about eleven forenoon. Dandelion flowers early in the ſpring, and again in the autumn. It expands at five or ſix in the morning, and cloſes them at eight or nine forenoon. Narrow-leafed buſhy hawkweed, expands about ſix in the morning, and cloſes about five afternoon. Succory-leafed-mountain-hawkweed, has its flowers expanded from ſix in the morning

A plant has a power of directing its roots for procuring food. A quantity of fine compost for flowers, happened to be laid at the root of a full grown

morning till five afternoon. The garden-hawkweed with deep purple flowers, expands from six or seven in the morning till three or four afternoon. The tree sow-thistle, common in corn-fields, flowers in June, July and August, expands about six or seven in the morning, and closes between 11 and 12 forenoon. The other species of the sowthistle, follow nearly the same course. Garden-lettuce expands about seven in the morning, and closes about ten forenoon. Hawkweed flowers in July or August. It expands about seven in the morning, and keeps expanded till about three in the afternoon. Bushy-hawkweed with broad rough leaves, flowers June and July; is expanded from about seven in the morning till one or two afternoon. Branched-spiderwort with a small flower, expands about seven in the morning, and closes between three and four afternoon. White water-lilly grows in rivers, ponds, and ditches; and the flowers lye on the surface of the water. At their time of expansion, about seven in the morning, the stalk is erected, and the flowers raised above the surface of the water. In this situation, it continues till about four in the afternoon, when the flowers sink to the surface of the water and close. Marygold with indented leaves, has its flowers expanded from seven in the morning till three or four afternoon. Linnæus observes of this plant, that if its flowers expand later than their usual time, it will most assuredly rain that day. The male pimpernel, flowers in June, and continues to flower three months: is expands about eight in the morning, and closes not till past noon. The blue-flowered pimpernel with narrow leaves, observes nearly the same time. The proliferous pink expands its flowers about

grown elm; where it lay neglected three or four years. When moved, in order to be carried off, there appeared a net-work of elm-fibres spread through the whole heap. No fibres had before appeared at the surface of the ground. The red whortleberry, a low evergreen plant, grows naturally on the top of our highest hills, among stones and gravel. This shrub was planted as an edging to a rich border, under a fruit-wall. In two or three years, it over-ran the adjoining deep-laid gravel-walk; and seemed to fly from the border, in which not a single runner appeared. Were our London aldermen equally temperate, they might partake of turtle and venison with safety. An effort to come at food in a bad situation, is extremely remarkable in the following instance. Among the ruins of Newabbey, formerly a monastery in Galloway, there grows on the top of a wall, a plane-tree about twenty feet high. Straitened for nourishment in that barren situation, it several years ago directed roots

about eight in the morning, and closes them about one afternoon. The flowers of wild succory, open about eight forenoon, and keep expanded till about four afternoon. Wild marygold has its flowers expanded from nine in the morning till three afternoon. The purple-spurry flowers in June, expands between nine and ten in the morning, and closes between two and three afternoon. Common purslain expands about nine or ten in the morning, and closes an hour after. The lesser water plantain opens its flowers about noon.

down the ſide of the wall, till they reached the ground ten feet below; and now, the nouriſhment it afforded to theſe roots during the time of their deſcending, is amply repaid, having every year ſince that time made vigorous ſhoots. From the top of the wall to the ſurface of the earth, theſe roots have not thrown out a ſingle fibre; but are now united into a pretty thick root.

Plants, when forced from their natural poſition, are endowed with a power to reſtore themſelves. A hope-plant twiſting round a ſtick, directs its courſe from ſouth to weſt as the ſun does. Untwiſt it, and tie it in the oppoſite direction: it dies. Leave it looſe in the wrong direction: it recovers its natural direction in a ſingle night. The leaves of all trees and vegetables, have an upper and an under ſurface which never vary. Twiſt a branch ſo as to invert its leaves, and fix it in that poſition. If left in any degree looſe, it untwiſts itſelf gradually, till the leaves be reſtored to their natural poſition. What better can an animal do for its well-fare? A root of a tree, meeting with a ditch in its progreſs, is laid open to the air. What follows? it alters its courſe like a rational being, dips into the ground, ſurrounds the ditch, riſes on the oppoſite ſide to its wonted diſtance from the ſurface, and then proceeds in its original direction. Lay a wet ſpunge near a root laid open to the air: the root will direct its courſe

courſe to the ſpunge. Change the place of the ſpunge: the root varies its direction.

Such animals as are naturally weak, exert their ſelf-motive power to remedy that defect, by joining in ſociety. Plants are not capable of ſociety; but ſeveral of them ſupply their natural weakneſs, by exerting their ſelf-motive power in a manner that would do honour to an animal. The œconomy of ſcandent plants is in that reſpect admirable. Obſerve how they direct their courſe to any thing that can ſupport them. Thruſt a pole into the ground, within a moderate diſtance: a ſcandent plant directs its courſe to the pole, lays hold of it, and riſes on it to its natural height. A honeyſuckle proceeds in its courſe, till it be too long for ſupporting its weight; and then ſtrengthens itſelf by ſhooting into a ſpiral. If it meet with another plant of the ſame kind, they coaleſce for mutual ſupport; the one ſcrewing to the right, the other to the left. If a honeyſuckle twig meet with a dead branch, it ſcrews from the right to the left. The claſpers of briony ſhoot in a ſpiral, and lay hold of whatever comes in their way for ſupport. If after completing a ſpiral of three rounds they meet with nothing, they try again by altering their courſe.

Nature has alſo provided a remedy for trees that grow too faſt in a fruitful ſoil. Some form the upper part of the weak and tender ſtem into a ſort of ſcrew; which is ſtronger than a ſtreight line.

line. This among others is the caſe of the larix. A tree bent by too faſt growing, puſhes out all its lateral branches on the convex ſide, in a direction between perpendicular and horizontal, as if it were expanding wings to raiſe itſelf up. There are at Kames, elms that when twenty feet high were bent down by overgrowing, the top almoſt touching the ground. In that poſition they continued ſeveral years, till they were raiſed by lateral branches as above deſcribed; and they are now perfectly erect.

The œconomy of ſome water-plants is ſingular. As the *farina fœcundans* cannot operate under water, a water-lillie, be the water deep or ſhallow, puſhes up its flower-ſtems till they reach the ſurface, and then flowers in open air.

The compariſon between plants and animals, may be carried a great way farther. There are powers in every animal, to ſtruggle for health by expelling diſeaſes. All that a ſurgeon can do in the caſe of a broken bone, is to reſtore it to its natural poſition. Nature performs the cure, by pouring into the broken part a liquid matter, which, hardening into bone, unites the parts firmly together. Sydenham, prince of phyſicians, defines a fever to be an effort of nature to throw out of the body what is noxious. The proviſion of nature for reſtoring a maimed animal, is remarkable in the lobſter and crab. The feeling of theſe animals is at the tip of their claws. When the

the tip of a claw is bruiſed or broken, the whole claw falls off, and another in its ſtead quickly arrives at maturity *. Are not yawning, ſtretching, ſighing, weeping, efforts of nature to throw off a burden? There are ſimilar powers in plants to remedy what is noxious. A wound in a tree is cured like a wound in an animal: the ſeparated parts unite; and the tree is covered with bark as formerly. If part of a branch or of a root be cut off, the want is ſupplied by a number of ſmall ſhoots, iſſuing from the place where the cut was made.

The foregoing facts exhibit a ſtrong reſemblance between plants and animals, with reſpect to the ſelf-motive power. The motion of the heart in animals, of the arteries, of the inteſtines, of the lungs, cannot be explained by any known law of mechaniſm; and as little, the ſpringing of the ſeed in plants, the oſcillatory motion of the ſap, the production of leaves, flowers, fruit, &c. Theſe various effects proceed from a ſelf-motive power in plants, as well as in animals; and by that power chiefly, are organized bodies diſtinguiſhed

* Monſieur Bonnet is ſtrangely puzzled to account for this fact He ſuppoſes that numberleſs embryos of every portion of every claw of a lobſter were originally created; that every lobſter, is full of ſuch embryos, ſo artificially placed, as that when part of a claw is broken off, an embryo correſponding to that part is at hand, which is put in motion in order to repair the loſs.

guiſhed from brute-matter. A plant exerts this power without conſciouſneſs, becauſe conſciouſneſs is not the property of any plant. But ſo far the reſemblance holds, that an animal, though endowed with conſciouſneſs, exerts the ſame power blindly, without being conſcious of the exertion. The power is in both exerted uniformly without interruption; and each individual may be conſidered as a ſort of *perpetuum mobile*.

The power that a plant or an animal has to remedy any diſorder or hurt, differs from the power of carrying on life, in the following particular, that it is quieſcent till there be occaſion to exert it: it is exerted by circumſtances deſtructive to the health of the plant or animal; and independent of ſuch circumſtances, it would contradict the beauty and order of nature that it ſhould be exerted.

There are other powers in animals termed inſtinctive, by which they act blindly without any view to conſequences: hunger prompts them to eat, and cold to take ſhelter, without reflection or foreſight. Inſtinctive actions differ from thoſe above mentioned, being attended with conſciouſneſs, though not with foreſight: a duckling, even where hatched by a hen, goes inſtinctively into water: it knows where it is going; but knows not for what end. To inſtinctive powers in animals, there are no reſembling powers in plants: we we admit not in plants any knowledge or conſciouſneſs. Far leſs do plants exert any actions that reſemble voluntary actions in animals.

After much labour beſtowed on botany, and many volumes compoſed on that ſubject, it appears very little advanced above infancy: no other ſcience has made ſo ſlow a progreſs. I praiſe the diligence of our botaniſts: ſome of them have great merit. But, as far as I underſtand, their ſtudy has been moſtly confined to give names to plants, and to diſtribute them into claſſes; not by diſtinguiſhing their powers and properties, but by certain viſible marks. This is an excellent preparation for compoſing a dictionary: but it leaves us in the dark as to the higher parts of the ſcience, ſuch as are the moſt proper to engage a thinking and rational mind. No perſon who has given attention to the conduct of Providence, can entertain any doubt, that the powers and properties of plants are given for beneficial purpoſes. Have we not reaſon to hope, that theſe purpoſes will be unfolded, when botaniſts, tired of dictionary-making, ſhall ſoar higher in their inquiries. Then will botany become an intereſting ſcience, not inferior to any other in dignity and importance: then ſhall we have occaſion to admire more and more the wiſdom of the creation. How pleaſant to have it obſerved, that the humbleſt plant is framed with no leſs ſkill, than the moſt elevated animal!

So much upon a comparifon between plants and animals with refpect to motion. Another comparifon occurs, no lefs curious, and ftill more interefting, that the external frame is nicely adjufted to the internal, fo as to accomplifh in perfection the ends of Providence. No one who has ftudied natural hiftory, but muft be fenfible of this agreement in the animal creation. How well adapted are the claws of a lion and the talons of an eagle, to their rapacious nature. What fort of figure would an innocent lamb make, or timid dove, with fuch arms! The fhape of a fifh is vifibly contrived for moving in water: how abfurd would the animal be, if it had an averfion to that element. A duckling waddles by inftinct to the firft water it fees; for which it is fitted by its oily feathers: fuch an inftinct in a chicken, would be highly incongruous. The hoof of a horfe correfponds to his fhallow underftanding: fingers would be inconfiftent with the ufe that nature intends him for. Without fingers a man would be a miferable creature: he would always be contriving, but without power to execute *.

This truth would be equally evident in plants, were their nature and qualities as well known as of animals. A plant is an organized being, as well as an animal: if the external frame of the latter

* Wonderfully fhallow is the reflection of Helvetius, that the only excellency of a man above a horfe, is his having fingers.

latter be adapted to its internal frame, can we doubt of the ſame œconomy with reſpect to the former? In whatever manner particles of matter are formed into an organic body, of which we know nothing, one thing is certain, that the organic body acquires a nature very different from that of its conſtituent particles, and alſo new powers qualifying it for acting according to its deſtination. The power of gravity, of reſiſtance, of continuing motion, eſſential to matter in general, will never by any combination produce any thing but motion; but the power of producing a body ſimilar to itſelf, inherent in all organic bodies, is far ſuperior to the powers mentioned; and therefore muſt be a new power added in the formation of every organic body.

As plants were originally created of many ſpecies, each ſpecies has powers peculiar to itſelf, which preſerve the different ſpecies diſtinct, and conſequently preſerve uniformity among the individuals of the ſame ſpecies. Theſe powers, varioully modified in every different ſpecies, are exerted in the propagation of new plants, with leaves, flowers, ſeed, *&c.* peculiar to each ſpecies. And as a perfect agreement between the external and internal frame of plants, as well as of animals, is undoubtedly the plan of nature, incapable of defect or overſight; it may be taken for granted, that each external part contributes to the well-being of the plant, and that any alteration would be hurtful: to exchange, for example, the leaves of an oak and an aſh, would be prejudicial to both, perhaps deſtructive. Were we acquainted with the nature of different plants, we ſhould be able to account for the difference of ſize, of leaves, of roots, of colour, and of ſeed. We ſhould alſo be able to explain why ſome plants ſpring early, ſome late; why ſome are adapted to a hot climate, ſome to a cold; why ſome thrive beſt in dry ſoil, ſome in wet; why ſome produce flowers before leaves; and why ſome never ſhed the leaf.

This ſpeculation opens a wide field for obſervation and experiment, that may worthily employ the moſt acute philoſophers. Why not then attempt to peep into the nature and conſtitution of plants? The beſt we can make of that ſubject, will, I am afraid, be but conjectural. But fair and rational conjectures, which we may hope for, will give ſome entertainment to the curious inquirer. If we deſpair of acquiring ſuch knowledge in the internal conſtruction of plants, as to explain all the differences above mentioned, we may at leaſt hope to diſcover facts that will illuſtrate the agreement between external and internal ſtructure. I venture to ſuggeſt an inſtance or two. Some plants draw moſt of their nouriſhment from the ſoil, ſome from the air. Do not ſmall leaves correſpond to the former, and large leaves to the latter? I have ſeen a houſe-leek growing vigourouſly oufly on a dry mud wall, excluding rain entirely from the roots. But it has thick leaves, and many in number, which fit it for drawing its nourifhment from the air. It is not the light, nor the fun, that makes plants grow erect, but the appointment of nature. A fcandent plant has a tendency to grow erect like other plants; but as it is too weak to ftand erect, it has tendrils or clafpers, to lay hold of any fupport within reach. Why do certain trees never fhed the leaf, even in this country? Is it not a rational conjecture, that they are fitted by nature to bear cold, and that the cold of this climate does not fufpend their power of drawing nourifhment all the year round?

Were this important branch of botany diligently ftudied, I fondly hope, that confiderable infight might be obtained into the nature of plants, and poffibly into their medicinal effects. By that ftudy, the natural hiftory of plants may become no lefs inftructive and entertaining, than that of animals.

ARTICLE V.

PROPAGATION OF PLANTS.

EQuivocal generation is by all philofophers exploded from animal life; but fome continue to hefitate with refpect to vegetables. Animals, fay they, wandering from place to place, can ftock the

the earth with their progeny; but plants are fixed to the earth they grow in. It is urged, that plants are never wanting where the ſoil is proper for them; that iſlands raiſed by a volcano at a diſtance from any ſhore, are ſoon covered with graſs; that muſhrooms and other organized bodies, ſpring from rotten ſtumps of trees, where they were never ſeen before; that various plants riſe on the foundation of old houſes, when cleared of the rubbiſh; and that upon liming or dunging, white clover ſprings up in the very central parts of a wide extended barren moor, though the ſeed of white clover has not wings to carry it to a diſtance.

To account for theſe ſingular facts, it is held, that both plants and animals were originally organized atoms or embryos, having all neceſſary parts in miniature; that the earth, the water, the air are full of ſuch atoms, which begin not to unfold themſelves into plants or animals, till they happen to meet with a proper *matrix* or *nidus*; that in their original ſtate, they are too minute for any of our ſenſes, but that they become viſible by expanſion.

What means were employed at the creation to cover the earth with plants, may be conjectured, but is far beyond the reach of evidence. It is to me a rational conjecture, that a number of plants and animals were originally created, and endued with proper powers of generation; and that from theſe, all the plants and animals exiſting in the world

world are defcended. In that belief, I cannot fubmit to organized atoms, becaufe there is no evidence of them, and becaufe they are unnecef-fary. To illuftrate this conjecture, I add the fub-ftance of a letter I had the pleafure to receive from an eminent naturalift *.

" The doctrine of equivocal generation was
" univerfally admitted, till about 130 years ago;
" not, however, fo much by the ancients, as
" by the half-enlightened moderns. They faw
" mites in cheefe; and myriads of flies and creep-
" ing things in a dunghill, or a putrid marfh.
" Ignorance of the natural hiftory of thefe ani-
" mals, made way for conjecturing that they
" were mere fpontaneous productions, the effect,
" not of generation, but of corruption. This
" doctrine indeed was confined to thefe poor in-
" fects, and never was extended to a lion or a
" horfe. They did not advert, that to form a
" maggot and an elephant, require equal power
" and wifdom. The fame diftinction was car-
" ried into the vegetable kingdom. Becaufe
" no feed appeared to the naked eye in a fern, a
" mufhroom, or in any of the mofs-tribe, it was
" afferted, that none exifted; and while the oak
" and the laurel were dignified with generative
" faculties, thefe humble plants were vilified as
" the progeny of putrefaction. Equivocal gene-
" ration became thus an afylum for ignorance.

* Dr Walker minifter of Moffat.

" I

"I am clear to banifh equivocal generation
"from vegetables, as well as from animals; and
"I boldly maintain as a fundamental truth in na-
"ture, *omne vivum ex ovo.* By the *ovum* in
"vegetables, I mean, a feed, or any part of a
"plant that contains a bud, or is capable of
"forming it. They are in effect the fame; be-
"caufe every bud, as well as every feed, con-
"tains the embryo of a future plant. I know of
"no other way by which plants are propagated,
"but by feeds, fuckers, and layers. The laft is
"imitated by art, in cuttings, grafting, and ino-
"culation. Some late experiments are mention-
"ed of propagating trees by planting their leaves;
"but I do not believe it.

"Plants, it is true, are deftitute of locomo-
"tion; and by means of fuckers and layers, they
"can only cover contiguous fpots. But nume-
"rous and wonderful are the expedients, practi-
"fed by nature to diffeminate plants. Some feed-
"veffels burft with an explofive force, and throw
"the feed to a diftance. This is the cafe of our
"whin: did the feeds fall perpendicularly down,
"they would be fuffocated in the heart of an im-
"penetrable bufh. Some feed-veffels open not
"till wet with rain; but the feeds are found to
"fuffer by drought and to require immediate
"moifture when fown. The afh and the plane
"have heavy feeds; but thefe feeds are fupplied
"with wings: a gale of wind caries them from
"their

" their lofty ſituation to a diſtance, and they re-
" main on the tree till the gale comes. The ſeeds
" of humble plants, that they may riſe and re-
" move, ſpread more ſail to the wind : the thiſtle
" ſpreads his beard ; and away he travels to fix
" his reſidence in remote parts. A plant of this
" kind, *Erigeron Canadenſe*, was imported from
" Canada about one hundred years ago, into the
" Paris garden. It is now ſpread as a wild plant
" over France, Holland, Germany, Italy, and
" it is ſaid over Sicily. It is ſpread to ſuch a de-
" gree over the ſouth of England, as to be in-
" liſted among the indigenous plants. Some
" ſeeds, ſuch as our clot-bur, are of an adheſive
" nature : they lay hold of animals that come
" near them, and are ſpread far and near.

" Many other agents are employed by nature,
" to ſtock the earth with plants. The ſea and
" rivers waft more ſeeds than they do ſails, from
" one part of the world to another. I have
" found ſeeds caſt aſhore in the Hebrides, that
" had been dropt accidentally into the ſea among
" the Weſt-India iſlands. The iſland of Aſcen-
" ſion, is the droſs of a volcano of a recent date.
" Its immenſe diſtance from land, renders its ac-
" quiſition of ſeeds difficult and precarious. I
" know but of two ways for ſupplying it with
" ſeeds, one by the waters of the ocean, the o-
" ther by birds. By one or other of theſe ways,
" it has got poſſeſſion of three ſpecies of plants

" and

" and only three; a ſingularity no where elſe
" known *."

" The animal creation is ſupported by the ve-
" getable; but in return, vegetables owe much
" of their progreſs to birds and graminivorous
" quadrupeds, which are prime agents in the diſ-
" ſemination of plants. Many birds live on fruits
" and berries: the pulp is their aliment; and
" they diſcharge the ſeeds unimpared, and ſpread
" them every where. Theſe ſeeds are heavy,
" and unprovided with any *apparatus* for flight;
" but the birds ſerve them for wings. Hence
" may be ſeen plantations of holly, yew, white-
" beam, rowan, ſpindletree, hawthorn, and ju-
" niper, formed by the birds of the air, upon im-
" pending cliffs and inacceſſible precipices. Be-
" cauſe the miſſelto grows upon trees, and has no
" flower that can be perceived; it was reckoned
" formerly a product of equivocal generation.
" It was concluded, that its large, round, heavy,
" berries, were not the ſeeds of the plant, be-
" cauſe they might fall to the ground, but never
" could mount up into trees. No berries are
" more palatable to birds of the thruſh kind; and
" it is they who plant them on high and diſtant
" trees.

* In no ſuch iſland was there ever found an animal that was not imported. And why ſhould we admit ſpontaneous generation to be more poſſible in a plant than in an animal?

"trees. It is extremely remarkable, that the ve-
"getating power of ſeeds, inſtead of being im-
"paired in the ſtomach of birds, ſeems to be
"fortified. The ſeeds of the magnolia, import-
"ed from America, commonly refuſe to vegetate
"under the management of the moſt ſkilful gar-
"deners. But I have heard, that theſe ſeeds,
"when voided by turkies, never fail to grow.
"It is well known, that the dung of domeſtic
"animals, while it fertilizes a garden, fills it
"with weeds. It approaches to a miracle, that
"ſeeds ſhould withſtand the power of animal di-
"geſtion which no other vegetable ſubſtance
"can do. Here is a meaſure laid down by Pro-
"vidence for the preſervation and diſſemination
"of ſeeds, that I cannot reflect upon without
"wonder.

"In order to fill the earth with plants, any o-
"ther method except by ſeeds, ſuckers, and lay-
"ers, appears to me unneceſſary, and therefore
"improbable. Farewell then to equivocal gene-
"ration. I can ſcarce write of it, without be-
"ing a little ruffled; ſo ill it correſponds with the
"more auguſt and comfortable ideas of creation,
"which have made a principal article of happi-
"neſs in my life."

So far my correſpondent. I join heartily with him in his concluſion, that the known means for ſtoring the earth with plants, which are conſpicuous marks of deſigning wiſdom, are in all ap-

 pearance

pearance ſo completely adequate, that to ſearch for unknown means, ſeems to be an idle attempt.

Having thus reſtored the plan of nature, which in the ſimpleſt manner employs ſeeds as the chief means for propagating plants, we proceed to conſider how ſeeds are formed. Many philoſophers, holding it to be incredible, that a plant, or even an animal, ſhould be endowed with a power to produce its own likeneſs, have embraced an opinion, that all the plants and animals that ever exiſted, or that ever can exiſt, were formed originally, not plants or animals, but embryos of thoſe incloſed in an egg or ſeed, which when depoſited in a proper *nidus* or *matrix*, grow up to a plant or animal, and then decay. And to account for future generations, it is held, that every embryo contains within it ſmaller embryos without end, like cups of different ſizes caſed one within another. Theſe philoſophers muſt go ſtill farther. To account for each ſeed producing a tree, and that tree producing ſeed, it muſt alſo be held, that the embryo incloſed in a ſeed contains ſmaller embryos decreaſing in ſize without end; and that each of theſe ſmaller embryos contains another ſeries of decreaſing embryos, alſo without end. Here are infinites upon infinites, ſtill without end. To avoid the intricacy of infinites upon infinites, ſome philoſophers have varied the ſyſtem a little, with a view to render it, as they think, more ſimple, by recurring to organized atoms, exploded above.

above *. But this ſyſtem, not to repeat what is ſaid againſt it above, is only in appearance more ſimple: it reſolves into infinites upon infinites like the former, and is in reality no leſs intricate. Take any of the ſuppoſed embryos, hovering in air, ſwimming in water, or fixed in earth: give it a proper *matrix*, and let it become a tree with ſeed. As each ſeed may produce a tree, and each tree produce ſeed which may alſo become trees, it is manifeſt, that an infinite number of embryos muſt have been contained in the firſt embryo, and an infinite number in every one of that infinite number.

That every ſeed contains an embryo-plant, is a valuable diſcovery in natural hiſtory; but that there is a decreaſing ſeries of embryos within every ſeed, is a mere conceit, aſſumed without the leaſt appearance of truth. So far is it from holding true that plants within plants ſubſiſt in a ſeed without end, that even the ſingle plant it contains is there in a very imperfect ſtate. The plume and radicle alone ſubſiſt in it; and the other parts are produced in the courſe of growing. But let us give way to the ſuppoſition of an infinite ſeries, to ſee what can be made of it. Writers ſtop ſhort and leave the reader in the dark, preciſely where he needs light the moſt. A ſeed is laid in earth: by what mechanical power is vegetation

* See Bonnet upon organized bodies.

vegetation produced and continued during the life of the plant? And by what mechanical power does motion commence in the fœtus of an animal, and the blood circulate? When a ſeed happens to be inverted in the ground, with its radicle above, and its plume below; what is the mechanic power that makes them wreathe about the ſeed till the radicle gets into earth and the plume into air*? Unleſs theſe particulars can be accounted for mechanically, an embryo muſt be held a pure viſion. A power muſt be admitted even in the ſmalleſt embryo, to expand itſelf into a plant or animal, where it happens upon a proper nidus. And yet the admiſſion of that power deſtroys the hypotheſis, root and branch. A ſeed thrown into the ground would reſt there for ever, were it not endued with a power to begin vegetation, and to continue it. It grows into a tree: why may not that tree be endued with a power to form its own ſeed? If ſo, there is no neceſſity to go farther back: organized atoms or embryos muſt vaniſh, becauſe their is no uſe for them. Power in a tree to form its ſeeds, is no more extraordinary than that of ſucking juices from

* To aſcend and deſcend is not the ultimate view in theſe two parts, but to get into the air and earth. As ſeeds are generally depoſited on or near the ſurface of the ground, the plume aſcends and the radicle deſcends. But place a ſeed in an inverted flower-pot with earth in it: the radicle aſcends and the plume deſcends: the firſt purſues his road into the earth; and the other into the air.

from the earth, and converting them into its own ſubſtance, a power that every plant is admitted to have. And if plants have power to form ſeed, there ſurely can be no heſitation in aſcribing the ſame power to animals. Can any thing be more ſimple, or more agreeable to the analogy of nature, than that the Almighty, who created plants and animals, ſhould endue them with a power to propagate their kind? Are we not informed of this by eye-ſight; and can any ſolid argument be urged againſt what we ſee? Thus the operations of nature, when underſtood, turn out no leſs illuſtrious for their ſimplicity than for their extenſive effects.

I ſhall cloſe this eſſay with a paſſage of another letter from my correſpondent above-mentioned. "As for the doctrine of organized atoms diffuſed "through the unverſe in order to be converted "into animals and vegetables, it is not counte-"nanced by any thing within the ſphere of my "knowledge. No facts are adduced, nor do I "recollect any, to ſupport it. I adhere more and "more to this plain truth, that all plants and a-"nimals are propagated by ſeeds, or analogous "organizations; which are formed out of unor-"ganized matter, by the power of the vital prin-"ciple of plants and animals, in the way of ſecre-"tion. By analogous organizations, I mean a "bud of a tree, a ſection of a polypus, and ſuch-"like organized parts, that are capable like ſeeds of

" of growing up into a complete plant or ani-
" mal.

" The ſecretory power of plants and animals is
" indeed a wonder. A lyncean anatomiſt, with
" his great magnifiers, cannot penetrate the dark-
" neſs in which it is involved. The tranſmuta-
" tion of matter by animal and vegetable ſecre-
" tion, is obvious to every eye. By what means
" it is performed, ſeems to be that high legerde-
" main which nature will never reveal. But if
" by this power bread and water can be changed
" into fleſh and blood, into bones and ſinews,
" into the Argus' eye on the peacock's tail; if by
" this power ſimple water can be converted into
" the hardeſt wood, into aromatic flowers and
" rich fruits; I then ceaſe to wonder, that the
" ſame water ſhould be converted into a ſeed, ca-
" pable of unfolding itſelf into a future plant. I
" require no aid from vagrant organized atoms:
" I ſee no aſſiſtance they can afford. I diſlike a
" hypotheſis that appears not to have any foun-
" dation in truth, or even in probability."

F I N I S.

Printed in the United States
By Bookmasters